WHALES
DOLPHINS
—AND—
PORPOISES

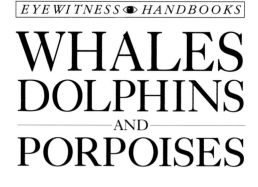

WHALES
DOLPHINS
—AND—
PORPOISES

MARK CARWARDINE

Illustrated by
MARTIN CAMM

Editorial Consultants

DR. PETER EVANS
(Sea Watch Foundation, University of Oxford, UK)

MASON WEINRICH
(Cetacean Research Unit, Gloucester, Massachusetts)

A DK PUBLISHING BOOK

Project Editor Polly Boyd
Designer Sharon Moore
Assisting Editor Lucinda Hawksley
U.S. Editor Charles A. Wills
Series Editor Jonathan Metcalf
Series Art Editor Peter Cross
Production Controller Meryl Silbert

First American Edition, 1995
10 9 8 7 6 5 4 3

Published in the United States by DK Publishing, Inc.
95 Madison Avenue, New York, NY 10016

Library of Congress Cataloging-in-Publication Data

Carwardine, Mark.
    Whales, dolphins, and porpoises / by Mark Carwardine. -- 1st
American ed.
        p.   cm. -- (Eyewitness handbooks)
    Includes index.
    ISBN 1-56458-621-9 (hard). -- ISBN 1-56458-620-0 (flexi)
    1. Cetacea. 2. Cetacea--Pictorial works. 3. Cetacea-
-Identification. I. Title. II. Series.
QL737, C4C28 1995   94-33301
599.5--dc20   CIP

Computer page makeup by Sharon Moore
Text film output by The Right Type, Great Britain
Map film output by Pressdata, Great Britain

Reproduced by Dot Gradation, Great Britain
Printed and bound in Singapore by Kyodo Printing Co.

CONTENTS

AUTHOR'S INTRODUCTION

Watching whales, dolphins, and porpoises in the wild is probably the ultimate wildlife experience. Who can remain untouched by a 30-ton Humpback Whale launching itself high into the air, the sheer size of a Blue Whale, or the drama of a school of Common Dolphins riding the bow waves of a boat? Whale watching is now one of the fastest-growing tourist activities in the world, involving over 40 different countries and attracting more than four million people every year.

WHALES, dolphins, and porpoises are elusive creatures, spending a major portion of their lives underwater or in remote areas far out to sea; as a result, they can be extremely difficult to study. Our understanding of their distribution, behavior, and other aspects of their lives is changing all the time, as new information comes to light. We do not even know how many species exist: new species are still being discovered, and there are constant discussions about whether others should be split into two or more different kinds.

For the purposes of this book, we recognize 79 species. Martin Camm, the illustrator, and I are very fortunate in having seen a great many of these over a combined 30 years of whale watching, but we have by no means seen them all; indeed, there are species that no one has positively identified alive. Therefore, to produce this book, we have relied on many different sources of information: not just our own field notes but also a huge number of photographs, videos, books, scientific papers, articles, and, of course, conversations with experienced colleagues and friends.

We have included a certain amount of natural history information but, as this is already available in many other books, our priority has been to produce a more practical field guide. Even then, we have had to be fairly selective in our coverage, simply because there is so much variation between individuals of the same species. We have endeavored to illustrate or describe the most commonly encountered variations –

HARBOR PORPOISE

ATLANTIC SPOTTED DOLPHIN

MAIN DISTINCTIONS
The main difference between whales, dolphins, and porpoises is their size. However, there is a great deal of overlap and there are also other, more specific, distinctions: porpoises, for example, are unique in having spade-shaped teeth, and most of the larger whales have so-called baleen plates and no teeth at all.

HUMPBACK WHALE

but do not be surprised if you see animals that do not do everything *exactly* as stated in the text, or animals that are not *identical* to the illustrations. We have chosen to use three-dimensional images, with highlights and a sense of movement, to show the animals as they are likely to be seen in real life.

Distribution maps are an essential part of any field guide. But, as many species are poorly known and because there are vast areas of ocean that have never been explored, many of the maps in this book are based on limited information. They should therefore, be viewed as broad representations of the approximate distribution of the species concerned. Wherever possible, new information that came to light after the maps had been produced has been incorporated into the accompanying text.

We hope that this *Eyewitness Handbook* will inspire you to find, recognize, watch, enjoy, and respect whales, dolphins, and porpoises in the wild.

MAKING CONTACT
This extraordinary encounter with a Gray Whale is made all the more poignant by the fact that, despite centuries of persecution by humans, many whales, dolphins, and porpoises so readily accept us as friends: Gray Whales, in particular, are so welcoming and inquisitive that it is often difficult to tell who is really watching whom.

▽ KAYAKING WITH A HUMPBACK WHALE
This is the kind of experience that changes people's lives. Kayaking is a wonderful way to watch whales, though it is crucial not to startle them by approaching them too quietly. It can be dangerous, but the whales themselves seem to be aware of their own size and strength and, if they are treated with respect, are usually careful to avoid any mishaps.

HOW THIS BOOK WORKS

FOLLOWING the Introduction and the Identification Key, the main part of the book is arranged according to the major families of cetaceans. Each family is described in an introductory feature (see right). The family sections are further divided into species entries. Every entry gives detailed information on the main characteristics, behavior, and distribution of a particular species. The teeth or baleen are illustrated and, in some cases, a typical sounding dive sequence is shown, as well as any other interesting features or variations. A typical species entry and typical family introduction are shown here.

SPECIES ENTRY

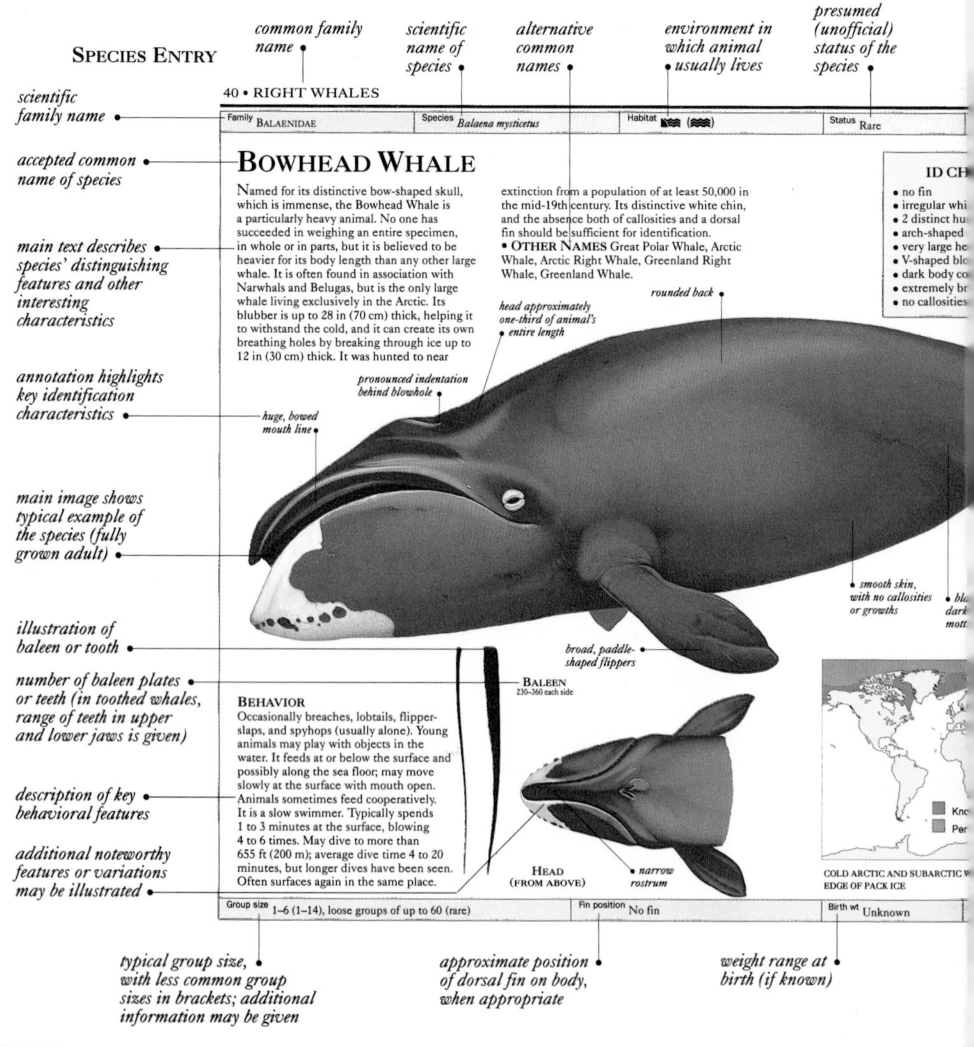

- common family name •
- scientific name of species •
- alternative common names •
- environment in which animal • usually lives
- presumed (unofficial) status of the species •

scientific family name •

accepted common • name of species

main text describes • species' distinguishing features and other interesting characteristics

annotation highlights key identification characteristics •

main image shows typical example of the species (fully grown adult) •

illustration of baleen or tooth •

number of baleen plates • or teeth (in toothed whales, range of teeth in upper and lower jaws is given)

description of key • behavioral features

additional noteworthy features or variations may be illustrated •

40 • RIGHT WHALES

Family BALAENIDAE	Species *Balaena mysticetus*	Habitat 〰〰 (〰〰)	Status Rare

BOWHEAD WHALE

Named for its distinctive bow-shaped skull, which is immense, the Bowhead Whale is a particularly heavy animal. No one has succeeded in weighing an entire specimen, in whole or in parts, but it is believed to be heavier for its body length than any other large whale. It is often found in association with Narwhals and Belugas, but is the only large whale living exclusively in the Arctic. Its blubber is up to 28 in (70 cm) thick, helping it to withstand the cold, and it can create its own breathing holes by breaking through ice up to 12 in (30 cm) thick. It was hunted to near extinction from a population of at least 50,000 in the mid-19th century. Its distinctive white chin, and the absence both of callosities and a dorsal fin should be sufficient for identification.
• **OTHER NAMES** Great Polar Whale, Arctic Whale, Arctic Right Whale, Greenland Right Whale, Greenland Whale.

ID CH
- no fin
- irregular whi
- 2 distinct hu
- arch-shaped
- very large he
- V-shaped blo
- dark body co
- extremely br
- no callosities

rounded back •

head approximately one-third of animal's entire length

pronounced indentation behind blowhole •

huge, bowed mouth line •

smooth skin, with no callosities or growths •

broad, paddle-shaped flippers

BALEEN
230–360 each side

BEHAVIOR
Occasionally breaches, lobtails, flipper-slaps, and syphops (usually alone). Young animals may play with objects in the water. It feeds at or below the surface and possibly along the sea floor; may move slowly at the surface with mouth open. Animals sometimes feed cooperatively. It is a slow swimmer. Typically spends 1 to 3 minutes at the surface, blowing 4 to 6 times. May dive to more than 655 ft (200 m); average dive time 4 to 20 minutes, but longer dives have been seen. Often surfaces again in the same place.

HEAD
(FROM ABOVE)

narrow rostrum •

COLD ARCTIC AND SUBARCTIC W
EDGE OF PACK ICE

Kno
Per

Group size 1–6 (1–14), loose groups of up to 60 (rare)	Fin position No fin	Birth wt Unknown

- typical group size, • with less common group sizes in brackets; additional information may be given
- approximate position • of dorsal fin on body, when appropriate
- weight range at • birth (if known)

FAMILY INTRODUCTION

common family • name or names

main text • introduces the family

main image • shows typical family member

examples of • baleen, teeth, or skull usually illustrated

estimated world population • size

major current threats to the species

RIGHT AND GRAY WHALES

T HREE VERY DIFFERENT families are included in this section: Balaenidae (Southern and Northern Right Whales and Bowhead Whale); Neobalaenidae (Pygmy Right Whale); and Eschrichtiidae (Gray Whale). The Southern and Pygmy Right Whales occur only in the southern hemisphere and the Bowhead Whale, Gray Whale, and Northern Right Whale are found only in the northern hemisphere. They all prefer temperate or polar waters. The Pygmy Right Whale has never been exploited by the whaling industry; however, the 4 larger species have suffered from tragic commercial overexploitation and all have, at some time, been close to extinction.

SPECIES IDENTIFICATION

PYGMY RIGHT WHALE (p.48) Smallest baleen whale, with a prominent dorsal fin.

GRAY WHALE (p.50) Arched head, and a low hump and "knuckles" instead of a dorsal fin.

BOWHEAD WHALE (p.40) White chin and a large head with no callosities, and no dorsal fin.

NORTHERN RIGHT WHALE (p.44) Large head covered with callosities, and no dorsal fin.

SOUTHERN RIGHT WHALE (p.44) Similar to the Northern Right Whale, but more common; lives only in the southern hemisphere.

strongly arched mouth line (except Gray Whale)

no true dorsal fin (except Pygmy Right Whale)

CALLOSITIES Southern and Northern Right Whales have growths, called callosities, above the eyes, by the blowholes, and on the chin, lower "lips," and rostrum.

BALEEN Bowhead Whales and Northern and Southern Right Whales have vast heads with long baleen plates; huge lower "lips" cover the plates when the mouth is shut.

"curtains" of baleen hang from rostrum only

FLIPPERS The flippers of the Bowhead Whale are considerably narrower and shorter (relative to body size) than those of Northern and Southern Right Whales.

SOUTHERN/NORTHERN RIGHT WHALES These whales show many physical characteristics common to all (or most) other members of these 3 families.

RIGHT WHALES   BOWHEAD WHALE

CHARACTERISTICS These 3 families share several features, but the Gray Whale and Pygmy Right Whale are exceptional in many ways. The Gray Whale is intermediate between the larger right whales and the rorquals (p.54); for example, it has a more streamlined body shape, yet has a hump instead of a true dorsal fin. The Pygmy Right Whale resembles its relatives in having an arched rostrum and no throat grooves, but it has a prominent dorsal fin.

PYGMY RIGHT WHALE The Pygmy Right Whale has a distinctive way of swimming; it makes an undulatory movement that is caused by waves of motion passing from its head to its flukes (increasing in amplitude toward the rear of the body).

flippers extended during slow swimming

flukes remain below surface when diving

shows little (if any) of dorsal fin or back at surface

S-shaped waves of movement along entire body

BOWHEAD WHALE SKULL

PYGMY RIGHT WHALE DIVE SEQUENCE

pulation 6,000–12,000   Threats

CKLIST

patch on chin s in profile uth

pointed tips •

Newborn 13¼–14¾ ft (4–4.5 m)
Adult 46–59 ft (14–18 m)

l flukes barnacles

distinct notch in middle

FLUKES

blue-black, dark gray, or own body color sometimes with large grayish patches

flukes' width can be almost half of total body length

MALE/FEMALE

flukes may have white trailing edges on upper sides

WHERE TO LOOK
There are 4 distinct populations: Davis Strait, Baffin Bay, northern Hudson Bay, and Foxe Basin; Bering, Chukchi, and Beaufort Seas; Sea of Okhotsk (may be part of Beaufort population); and the North Atlantic (virtually extinct). Not known whether these populations mix. Spends most of life at edge of Arctic pack ice (especially where over 70 percent ice cover). Short seasonal migrations related to formation and movement of ice (north in summer, south in winter).

range ent ice

ERS, RARELY FAR FROM

lt wt 60–100 tons   Diet

• weight range of fully grown adults

• main food (with less common food indicated in brackets, when appropriate)

• silhouettes show typical size of animal in relation to a human: all human silhouettes represent 8 ft (2.4 m)

• ranges of birth and adult lengths

• illustration of flukes' upper sides

• key features useful in identification (lists features characteristic of sex shown in main image)

• label indicates sex of animal in main illustration

• map shows assumed range of species and/or important sightings or strandings; where no key is given, blue shading shows known distribution only

• notes on distribution, migrations, and habitat

• typical or exceptional dive sequence (usually sounding dive)

lineup of family • members (in proportion) with brief description

△FAMILY INTRODUCTION
Each family or group of similar families has a short introduction describing its main characteristics.

KEY TO SYMBOLS
Data in the colored bands are concise to aid quick reference.

HABITAT
- inshore
- offshore
- riverine

DIET
- squid or octopus
- krill or other crustaceans
- other invertebrates
- fish
- mammals

THREATS
- entanglement in fishing nets
- pollution
- habitat destruction
- hunting/whaling
- human disturbance

WHAT IS A CETACEAN?

WHALES, DOLPHINS, and porpoises are known collectively as cetaceans, from the Latin *cetus* (a large sea animal) and the Greek *ketos* (sea monster). There are 79 species currently recognized, and it is very likely that we will discover more in the future. They come in a variety of shapes and sizes, ranging from tiny dolphins just over 39 in (1 m) long to the Blue Whale, which is typically 82 ft (25 m) long and is one of the largest animals ever to have lived on earth. Some cetaceans are long and slender, others are short and stocky. Some have huge dorsal fins, whereas others have no fins at all. While some are bright and conspicuous, others are drab and hard to see. They live in all the oceans and many major rivers of the world, from the warm waters of the equator to the cold waters of the poles.

TAIL
A cetacean swims with the help of powerful muscles in the rear third of the body. In smooth, measured movements, these force the animal's tail up and down to propel the body through the water.

FINS
Cetaceans have just a single dorsal fin, and even this is absent in some species. Most fish, however, have more than one fin: the Whale Shark, for example, has four in total.

WHALE

TAIL
A fish swims by moving its head from side to side, which sends "waves" down its body. These waves increase in intensity and finally reach the tail, which then swings from side to side. It is this movement that propels the animal through the water.

SHARK

SKIN
A shark is covered in thousands of rough, toothlike scales, but a cetacean's skin is smooth to touch.

ORIGINS OF CETACEANS

Cetaceans probably evolved from furry land mammals with four legs. The first real whalelike animals, called Archaeocetes, appeared about 50 million years ago. They were not the direct ancestors of modern cetaceans, but were probably very similar. There were many different kinds, ranging in length from about 6½ ft (2 m) to 69 ft (21 m), and they are believed to have lived in coastal swamps and shallow seas. They had torpedo-shaped bodies and their forelimbs had turned into paddles. Archaeocetes died out around 30 million years ago.

long, torpedo-shaped body

dorsal fin

long tail stock; hindlimbs have disappeared

paddles evolved from forelimbs

ANCIENT WHALE

DIET OF CETACEANS

Cetaceans eat a variety of food. Their choice depends on their size, whether or not they have teeth, and various other factors. Most of the larger whales feed on huge shoals of fish or tiny, shrimplike creatures such as krill, while dolphins and porpoises tend to catch individual fish or squid. Other, less common, prey items include octopuses, mollusks, polychaete worms, crabs, turtles, and marine mammals, including other cetaceans.

KRILL
Krill are tiny but full of protein. Their swarming habit makes them easy prey for large whales.

BLOWHOLES
A cetacean cannot extract oxygen from the water, so it rises to the surface at regular intervals to breathe air. It has special "blowholes," on the top of its head, instead of gills.

FISH OR CETACEAN?

At first glance, whales, dolphins, and porpoises resemble fish, particularly sharks. Both the Fin Whale and the Whale Shark shown here, for example, have remarkably similar body shapes and both have dorsal fins, flippers, and huge tails. In fact, the similarities are so striking that, for many years, whales and all other cetaceans were thought to be "spouting fish." However, they are mammals and more closely related to people: they are warm-blooded, breathe air, and give birth to live young. The best way to distinguish a cetacean from a fish at a glance is to look at the tail: a whale's tail is horizontal and moves up and down, while a fish's tail is vertical and moves from side to side.

EARS
Cetaceans do not have external ears, but tiny ear openings, located a little way behind each eye. They have an excellent sense of hearing and, unlike land mammals and fish, can tell the direction of sound underwater.

GILLS
Fish do not need to rise to the surface to breathe. With the help of their gills, they can take all the oxygen they need directly from the water.

GROWTH RINGS
It is possible to estimate the age of some species by studying the series of layers, rather like the growth rings in a tree, inside their teeth. Broadly speaking, one complete layer is equal to a year of growth.

FLIPPERS
Both fish and cetaceans have flippers, or pectoral fins. Shaped like paddles, these modified forelimbs are used primarily for twisting and turning. Flipper sizes, shapes, and colors vary considerably from one species of cetacean to another and, in some cases, from one individual to another.

MOTHER AND CALF

calf is lifted to surface by its mother

GIVING BIRTH

Like most mammals, whales, dolphins, and porpoises give birth to live young. They usually have just one baby at a time. It is born underwater, near the surface, and normally comes out tail first. The newborn calf is a little awkward initially and may have to be nudged toward the surface for its first breath by the mother or an "assistant." Few births have been seen in the wild.

ANATOMY OF A CETACEAN

THERE ARE two main types of whale: toothed whales, or Odontocetes, which possess teeth; and baleen whales, or Mysticetes, which do not. The toothed whales include the Narwhal and Beluga, all the dolphins and porpoises, sperm whales, and beaked whales; they feed mostly on fish, squid, and, in a few cases, marine mammals, and normally capture one animal at a time. The baleen whales include most of the larger whales, such as the rorquals, right whales, and the Gray Whale, and have baleen plates instead of teeth; their vast jaws enable them to catch thousands of shrimplike crustaceans or small fish at a time.

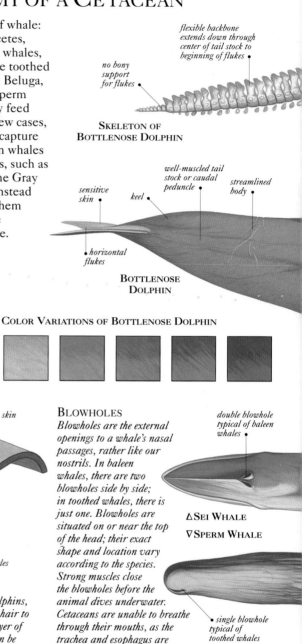

flexible backbone extends down through center of tail stock to beginning of flukes •

no bony support for flukes •

SKELETON OF BOTTLENOSE DOLPHIN

well-muscled tail stock or caudal peduncle •

streamlined body •

sensitive skin •

keel •

• horizontal flukes

BOTTLENOSE DOLPHIN

COLOR VARIATIONS

Most cetaceans have distinctive colors and body patterns, but these may vary between animals of the same species. The sexes are sometimes different, individuals may change color as they age, and there can be many geographical variations. Even animals of the same age, sex, and population may look dissimilar. Some color variations of the Bottlenose Dolphin are shown here.

COLOR VARIATIONS OF BOTTLENOSE DOLPHIN

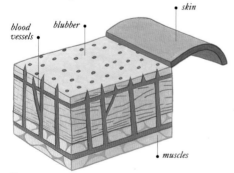

• skin

blood vessels •

blubber •

• muscles

BLUBBER

Unlike most other mammals, whales, dolphins, and porpoises do not have thick coats of hair to keep them warm. Instead, they have a layer of insulating fat, known as blubber; this can be as thick as 20 in (50 cm) in some species.

BLOWHOLES

Blowholes are the external openings to a whale's nasal passages, rather like our nostrils. In baleen whales, there are two blowholes side by side; in toothed whales, there is just one. Blowholes are situated on or near the top of the head; their exact shape and location vary according to the species. Strong muscles close the blowholes before the animal dives underwater. Cetaceans are unable to breathe through their mouths, as the trachea and esophagus are completely separate.

double blowhole typical of baleen whales •

△ SEI WHALE
▽ SPERM WHALE

• single blowhole typical of toothed whales

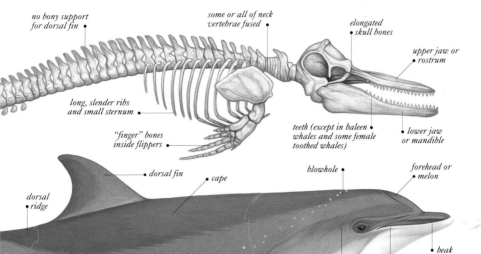

no bony support for dorsal fin

some or all of neck vertebrae fused

elongated skull bones

upper jaw or rostrum

long, slender ribs and small sternum

"finger" bones inside flippers

teeth (except in baleen whales and some female toothed whales)

lower jaw or mandible

dorsal fin

cape

blowhole

forehead or melon

dorsal ridge

beak or snout

mouth line

tiny opening instead of external ear

genital slit

mammary slits on either side of genital slit (female)

navel or umbilicus

belly

flippers or pectoral fins

AQUATIC MAMMALS

Cetaceans have shed most external traces of their terrestrial ancestry (see p.10) and are all supremely adapted to underwater life. Their body shape has become streamlined and they have lost most of their body hair, improving hydrodynamic efficiency; they have short, stiff necks, which are essential for swimming at high speed; their forelimbs have turned into flippers and their hindlimbs have disappeared; they have muscular tails providing a powerful means of propulsion; and their nostrils have moved to the top of the head for easy breathing at the surface.

There are also many other, less obvious adaptations. For example, they have excellent hearing, which compensates for a poor – or entirely lacking – sense of smell, and for the uncertainties of visibility underwater; they have a high tolerance to carbon dioxide, to help with lengthy dives, and are two or three times more efficient than land mammals at using oxygen in inhaled air; their rib cages are collapsible for deep diving; and they have layers of insulating fat to keep them warm.

BALEEN PLATES

Baleen whales have hundreds of comb-like baleen plates, or "whalebones," hanging down from their upper jaws. The plates overlap inside the mouth and have stiff hairs, which form a sieve to filter food out of the seawater.

CETACEAN BEHAVIOR

AFTER YEARS of studying dead animals washed ashore, or killed by whalers, researchers have learned a great deal about the anatomy and physiology of cetaceans but surprisingly little about their behavior. It is extremely difficult to study animals that spend most of their lives underwater, often far from land, but with recent technological advances and increased efforts to study cetaceans in the wild, we are now beginning to discover some of their secrets.

HUMPBACK FLIPPER RAISED IN THE AIR

BREACHING

Whales, dolphins, and some porpoises sometimes launch themselves into the air head first and fall back into the water with a splash. This is known as "breaching" and is undoubtedly the most spectacular surface activity. Often, it provides the only opportunity whale watchers have to see the entire animal.

Most species have been observed breaching at one time or another. The smaller cetaceans can leap very high and often do complete somersaults, twists, and turns before re-entering the water. Larger whales normally propel at least two-thirds of their bodies into the air, and their breaches end in a belly flop, or they turn to one side or onto their backs. Some also perform "head-slaps," which resemble breaches but involve lifting only the head and top part of the body

HUMPBACK FLIPPER HITTING THE SURFACE

△ FLIPPER-SLAPPING
Whales and dolphins sometimes roll over at the surface to slap their flippers onto the water with a splash – sometimes several times in a row – as shown in these two photographs of a Humpback Whale (see above). This action is known as "flipper-slapping," "flipper-flopping," or "pectoral-slapping." Humpback Whales sometimes also lie on their backs, waving both flippers in the air, before slapping them onto the surface of the water simultaneously.

FLUKING
When some whales and dolphins embark on a deep dive, or "sounding dive," they lift their tails into the air to help them thrust their bodies into a more

steeply angled descent to deeper waters. This is called "fluking." There are basically two kinds of fluking: a "fluke-up dive," when the flukes are brought high into the air, so that the undersides are out of the water before pounding them onto the surface. Many species breach several times in a row, and when one animal breaches, others may follow suit.

Humpback Whales have been known to breach over 200 times in a single display, both at their feeding grounds and breeding grounds; this is a phenomenal achievement, considering an average-sized Humpback weighs about as much as 400 people.

Breaching is still something of a mystery, although there are numerous possible explanations: it may be a courtship display, a form of signaling, a way to herd fish or dislodge parasites, a show of strength or a challenge, or it could be simply for fun. In fact, it probably has several of these functions.

◁ BREACHING
Breaches range from a full leap clear of the water to a more leisurely surge, in which only half the body emerges. It is not unusual for a Bryde's Whale to breach (see below), but Gray Whales, Humpback Whales, Sperm Whales, right whales, and many dolphins are better known for their breaching.

◁ LOBTAILING
"Lobtailing" describes the forceful slapping of the flukes against the water while most of the animal lies just under the surface; it is also known as "tail-slapping." Lobtails may be repeated many times in a row. A superficially similar form of behavior is "peduncle-slapping," or "tail-breaching"; this involves throwing the rear portion of the body out of the water and slapping it sideways onto the surface or on top of another whale. Tail breaches are similar to normal breaches, but are tail first instead of head first; in some species, they are believed to be a form of aggression.

visible, as shown in these photographs of a Sperm Whale; or a "fluke-down dive," when the flukes are brought clear of the water but remain turned down, hiding the undersides from view. When

watching a whale, observe whether or not it raises its flukes into the air at the start of a deep dive, and if it does, note what shape they are. Both are very helpful features for identification.

◁ IDENTIFYING THE BLOW
Experienced whale watchers can tell one species of whale from another, even from a considerable distance away, just by the character of the blow and with a brief glimpse of the animal's back. A right whale's blow consists of two separate columns of water vapor, for example, while in Blue Whales and Fin Whales these merge to form a single spout. This low, bushy blow, which is leaning backward slightly because of the wind, belongs to a Humpback Whale.

BLOW

One of the best ways of locating large whales in a vast expanse of sea is by their breath. This is known as a "blow" or "spout." The term refers both to the act of breathing – an explosive exhalation followed immediately by an inhalation – and to the cloud of water droplets produced above the animal's head when it breathes out.

The blow varies in height, shape, and visibility among species and, especially on calm days, can be very distinctive; however, in wind and rain the water droplets disperse more rapidly and the shape of the blow is likely to change.

No one really knows what makes a blow so visible. It probably includes water vapor that has condensed in the cold air, and a small amount of seawater that is trapped in the blowhole, but it may also contain a fine spray of mucus from inside the whale's lungs. In many smaller cetaceans, the blow is low and brief and, if it is visible at all, rarely has a distinctive shape.

△ BEACH-RUBBING
Whales, dolphins, and porpoises are tactile creatures. To the delight of whale watchers, some animals even enjoy having their noses scratched and rub themselves on the hulls of stationary ships. This Killer Whale is rubbing itself on pebbles in shallow water near the shore.

▽ SPYHOPPING
Many cetaceans periodically poke their heads above the surface of the water, perhaps to have a look around. Gray Whales, for example, slowly rise straight upward, until their eyes are just visible, and then may turn in a small circle before slipping back below the surface.

WAKE-RIDING▷

Swimming in the frothy wake of a boat or ship seems to be a favorite pastime for many dolphins, as well as some whales and porpoises. They surf in the waves, twist and turn or swim upside down in the bubbles and, like this Bottlenose Dolphin, often perform acrobatics.

PLAY

It is difficult to imagine that certain forms of behavior observed in whales and dolphins can be anything more than exuberant play. They chase one another, jump in the air, launch into bursts of erratic swimming, and twist and turn in the water. If they hear a passing boat or ship, they actively follow it, going out of their way to wake-ride or bow-ride. Many of them seem to enjoy the company of people, seals, sea turtles, and a variety of other species; they even play with pieces of seaweed, pebbles, and other objects in the sea, carrying them in their mouths or balancing them on their flippers. Inevitably, there are logical explanations for some of these activities, but there is little doubt that play for its own sake also has an important role in their lives. In young animals, for example, it forms part of the learning process, and in adults it may help to strengthen social bonds.

△BOW-RIDING

Many cetaceans, especially dolphins, frequently ride the bow waves of boats and ships. They jostle for the best position, where they can be pushed along in the water by the force of the wave. Some smaller cetaceans ride the bow waves of large whales in exactly the same way.

ECHOLOCATION

Most whales, dolphins, and porpoises are able to build up a "picture" of their surroundings with the help of sound. This is called echolocation. They make noises that bounce back from nearby objects and alert the animal that something else is in the water. Bats use a similar system to find their way around in the dark.

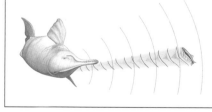

△LOGGING

A group of cetaceans, like these Short-finned Pilot Whales, may be seen at the surface, floating motionless, and all facing in the same direction. This is known as logging and is a form of rest.

How Cetaceans Are Studied

WHALES, dolphins, and porpoises are relatively difficult animals to study in the wild. Many species live in remote areas far out to sea, spend much of their lives underwater, and often show little of themselves when they rise to the surface to breathe. Some of them are also fairly shy and elusive and will avoid boats, making close encounters almost impossible. Several larger species divide their time between separate feeding and breeding areas, which are often hundreds or thousands of miles apart. Not surprisingly, for many years the only information we had about them all came from dead animals that had been washed ashore or killed by whalers or fishers. Nowadays, however, a number of interesting research techniques are being employed to study the animals in their natural habitats, and exciting new discoveries are being made all the time.

WATCHING CLOSELY
The more we discover about cetaceans – even those that are readily approachable, like this Bottlenose Dolphin – the more we realize there is to learn.

▽HUMPBACK WHALE DISTRIBUTION
This map shows the major feeding and breeding areas of Humpback Whales, with their possible spring and autumn migrations in between. It is the result of more than a century of research by dozens of scientists worldwide and continues to be adjusted as more information comes to light.

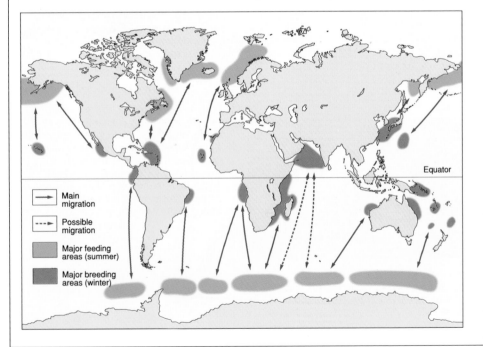

Equator

→ Main migration

--▶ Possible migration

■ Major feeding areas (summer)

■ Major breeding areas (winter)

IDENTIFYING INDIVIDUALS

Many cetacean research projects rely on the ability of scientists to identify and follow the daily activities of individual animals. Sometimes it is possible to recognize them by variations in the shape of their dorsal fin or by unique natural markings, such as those caused by scarring and pigmentation patterns. A selection of equipment is also available, ranging from depth-recorders to satellite transmitters, for more complex studies of cetaceans where direct observation is impossible.

SHORT-FINNED PILOT WHALES▷

Every Short-finned Pilot Whale has a unique dorsal fin: the animal's personal history is "engraved" on it in the form of scars, scratches, and nicks. These markings, combined with the actual shape of the fin (which can be anything from triangular to sickle-shaped and curled over at the top), make it relatively easy to tell one individual from another.

DISTINCTIVE NICKS   **HOOKED TIP**

"SALT"   **"DASH-DOT"**

◁HUMPBACK WHALES

It is possible to recognize individual Humpback Whales by the distinctive black and white markings on the undersides of their flukes. These range from pure white to jet black and include many variations in between. No two Humpbacks have exactly the same markings, just as no two people have exactly the same fingerprints; this means scientists can follow a particular animal's activities from one day or from one year to the next. Thousands of individual Humpbacks have been identified and studied in this way. These Humpbacks were photographed off the coast of New England, where they are well known to biologists in the area.

"SEAL"   **"HAWKSBILL"**

ARTIFICIAL TAGGING▷

Many cetaceans are difficult to observe closely at sea, and in some species it is almost impossible to tell one individual from another by their natural markings. In these circumstances, scientists sometimes use artificial devices known as tags to identify specific animals at a distance or even when they are completely out of sight. Satellite transmitters, like the one shown on this Dwarf Sperm Whale, are the most high-tech tags: they send signals up to a satellite orbiting the earth, which then beams them back down to powerful receiving stations on the ground.

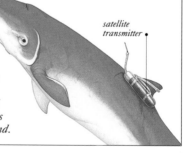

satellite transmitter •

CONSERVATION

A FEW CENTURIES AGO, there were probably more cetaceans in the sea than there are today. Whaling and other forms of hunting, incidental capture in fishing nets, competition with fisheries for food, human disturbance, habitat destruction, and marine pollution have all taken their toll. No cetacean has become extinct in modern times, but some species are now in serious trouble and others have all but disappeared from many of their former haunts.

WHALING

Commercial whaling began hundreds of years ago, but two relatively recent events precipitated an enormous increase in the killing worldwide: in 1864 a new harpoon was developed that could be fired from a cannon and would then explode inside the whale's body; and, in the early 1920s, floating factory ships were introduced that could do all the processing at sea. One by one, large whales were hunted almost to the point of extinction. A worldwide ban was agreed on in 1986, after tireless campaigning by conservation groups, but several hundred whales are still being killed every year. There is still some commercial whaling, in open defiance of the ban, but most of the killing occurs as a result of a serious loophole that allows any nation to issue its own permits to take whales for scientific research: the carcasses, however, are allowed to be processed for their meat and oil in the normal way.

HUNTING SMALLER SPECIES

Smaller whales and dolphins are also being hunted, especially in Japan and South America, with nets, rifles, and hand-held harpoons. They provide meat

△ MODERN WHALING
Hundreds of large whales are slaughtered every year, yet there is no humane way to kill them and no one knows if their severely reduced populations can withstand the continuing pressure.

◁ TRADITIONAL
WHALING
Coastal communities have killed whales for hundreds of years, using small boats and hand-held harpoons. In some parts of the world, a small number of whales are still hunted, under special license, by traditional methods.

MARINE POLLUTION

Untreated sewage, toxic chemicals, industrial waste, agricultural run-off, and a huge variety of other human-made pollutants enter the sea every day. But severe accidents, such as the 1989 Exxon Valdez *disaster in Alaska, in which oil spilled over 9,600 square miles (15,445 km²), can be especially catastrophic. This picture shows the 1993* Braer *disaster in Shetland, Scotland.*

ENTANGLEMENT IN NETS▷

Hundreds of thousands of cetaceans become entangled and drown in fishing nets every year, as increasingly destructive fishing methods are used to achieve maximum profits.

for human consumption and bait for crab fisheries; in some areas, where fish stocks are severely depleted because of over-exploitation, these cetaceans are blamed for the scarcity of fish and are "culled" to "protect" the rest of the stock. Again, conservationists have been campaigning for many years to stop the hunting.

FISHING

In many parts of the world, an increasing number of cetaceans become entangled and drown in fishing nets every year. The problem varies according to the type of fishing gear used, local customs, the species, and the fishing season. In some cases, a simple modification of either the nets or the management system can significantly reduce the incidental catch; in others, there may be no alternative but to ban the fishery. Most of the fisheries that kill dolphins do so accidentally, but tuna fishing is a notable exception; many fishermen set their nets around dolphins intentionally, knowing that they tend to swim with the tuna, as it is the cheapest and quickest way to catch the most fish.

CAPTIVITY

Bottlenose Dolphins, False Killer Whales, Killer Whales, Belugas, and Irrawaddy Dolphins are just some of the cetaceans that are caught in the wild to be kept in marine parks and zoos world-wide. In captivity, many animals die long before they reach old age.

STRANDING

EVERY YEAR, thousands of whales, dolphins, and porpoises are found stranded, alive and dead, on beaches all over the world. They may be alone or in groups, and while some animals are old or sick, many of them are young and appear to be perfectly healthy. This is a natural phenomenon and has been happening since time immemorial, but it still remains one of the great unsolved mysteries of the animal kingdom.

POSSIBLE CAUSES

Some strandings are easy to explain: the animals simply die at sea and are washed ashore with the tides and currents. But live strandings are more mysterious, and many theories have been put forward to explain their possible cause. One theory is that changes in the earth's magnetic field (see right) cause the animal to lose its sense of direction. Alternatively, an earthquake or storm could cause it to panic; a brain infection could cause disorientation; its sonar system may fail; or it may simply get lost or feel sick and need to rest. In mass strandings, the whole group may be in trouble in some way, or they may be following one individual that is ill or disoriented.

FINDING A STRANDED CETACEAN

In most cases, a stranded cetacean will be unable to return to the sea without help. If you find a stranded animal, check to see whether it is alive or dead: listen for the breathing (in some species there may be as much as 10 to 15 minutes between breaths) and see if the eyes move. If the animal is dead, inform the local police and do not touch the carcass. If it is alive, contact the police before trying to make the whale more comfortable. The guidelines opposite are very basic – it is always better to leave it to the experts if possible.

MAGNETIC FIELDS

Cetaceans may have an extra sense called biomagnetism, which enables them to detect variations in the earth's magnetic field. They may use the magnetic field, like a map, to navigate. The field is always changing so, occasionally, they could become confused and swim toward the shore.

COMMON VICTIM

Some whales are more susceptible to stranding than others. Pilot whales seem to strand more often than most: the social bonds between them are so strong that they are reluctant to desert one another; as a result, large numbers of them may strand together.

HOW TO DEAL WITH A STRANDED LIVE ANIMAL

WHAT TO DO

- Get expert help (via the local police) as quickly as possible.
- Keep the animal's skin moist.
- Erect a shelter to provide shade.
- Keep the flippers and flukes cool.
- Keep onlookers at a distance.
- Make as little noise as possible.
- Try to keep the animal upper side up.

DO NOT

- Stand very close to the tail or head.
- Push or pull on the flippers, flukes, or head.
- Cover the blowhole.
- Let either water or sand enter the blowhole.
- Apply sunblock lotion to the animal's skin.
- Touch the animal more than necessary.

MASS STRANDING
A stranded animal is susceptible to sunburn and overheating, even in cold weather, so it should be kept moist and cool, with wet towels or water, as shown with these False Killer Whales in Australia. It is vital to get expert help as quickly as possible. In some countries, it is illegal to give unauthorized first aid to a cetacean.

INFORMATION FROM STRANDINGS

For many years, the only information available on cetaceans was gleaned from animals killed by whalers and fishers, or from strandings. Even now, despite increasing work being undertaken on healthy animals at sea, some species have never been seen alive, and many others are almost impossible to distinguish at sea. An examination of a dead specimen may be necessary for a positive identification; even badly decomposed animals can often be identified with certainty.

△**COLOR CHANGES**
Many cetaceans change color after death, sometimes within a few hours, and therefore give a false impression of their true coloration. Normally, the change involves a substantial darkening. This example shows the change observed in a Cuvier's Beaked Whale within the first 24 hours of death.

LITTLE-KNOWN WHALE▽
Longman's Beaked Whale is known only from two weathered skulls, found on beaches in Australia and Somalia. There are no other records for this species, making it the least known of the world's cetaceans.

EXAMINING THE TEETH

Stranding provides the opportunity to examine a cetacean's teeth; in most species they are not visible at sea. The size, shape, number, and location of a cetacean's teeth can be very distinctive and helpful in identification. The teeth shown here are from a Fraser's Dolphin.

front tooth •

• *back tooth*

WHERE TO WATCH CETACEANS

WITH A LITTLE LUCK, it is possible to see whales, dolphins, and porpoises almost anywhere in the world. A walk along the coast, a short ferry crossing, or even a harbor cruise – all of these can provide opportunities for some excellent sightings. But many species tend to be plentiful only in particular areas and, even then, are present only at certain times of the year; so, without forward planning, it is possible to spend many fruitless hours simply staring at an empty sea. Most commercial trips have a fairly high success rate largely because they tend to concentrate on well-known cetacean populations and, of course, the trips are operated in the appropriate seasons. The map below shows many of the locations around the world where there is a better-than-average chance of exciting close encounters with a variety of whales, dolphins, or porpoises.

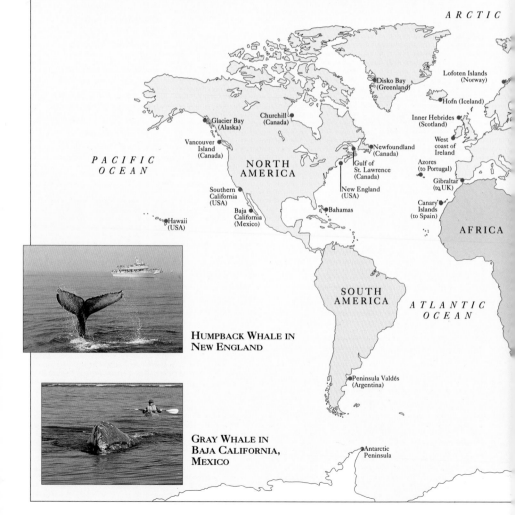

ARCTIC

Disko Bay (Greenland)

Lofoten Islands (Norway)

Hofn (Iceland)

Churchill (Canada)

Glacier Bay (Alaska)

Inner Hebrides (Scotland)

Vancouver Island (Canada)

Newfoundland (Canada)

West coast of Ireland

PACIFIC OCEAN

NORTH AMERICA

Gulf of St. Lawrence (Canada)

Azores (to Portugal)

Gibraltar (to UK)

New England (USA)

Southern California (USA)

Canary Islands (to Spain)

Baja California (Mexico)

Bahamas

Hawaii (USA)

AFRICA

HUMPBACK WHALE IN NEW ENGLAND

SOUTH AMERICA

ATLANTIC OCEAN

Peninsula Valdés (Argentina)

GRAY WHALE IN BAJA CALIFORNIA, MEXICO

Antarctic Peninsula

WATER TEMPERATURES

When describing where to find whales, dolphins, and porpoises, it is always useful to be able to name the areas of ocean in which they live. Since their distribution is often related to the surface water temperature, this is normally done using identifiable temperature zones, as shown on the map at the right. Although these zones tend to vary from season to season, they nonetheless provide us with an invaluable point of reference.

Tropical, Subtropical

Warm Temperate

Cold Temperate

Subarctic, Arctic / Subantarctic, Antarctic

Permanent Ice

Arctic Circle

Tropic of Cancer

0° Equator

Tropic of Capricorn

Antarctic Circle

TEMPERATURE ZONES

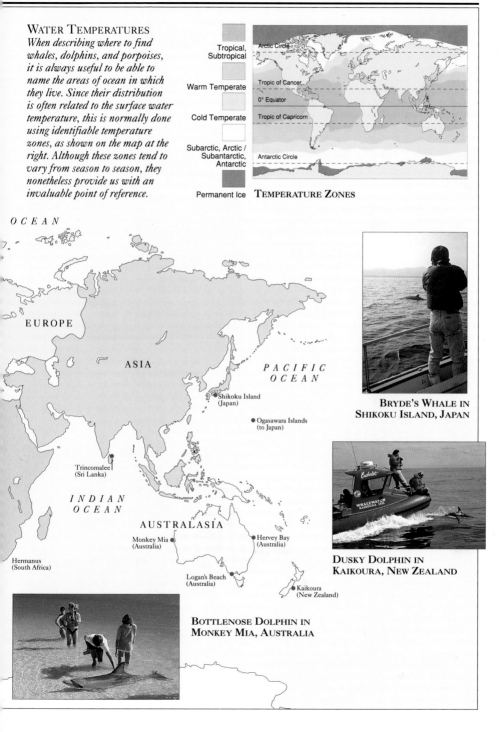

OCEAN

EUROPE

ASIA

PACIFIC OCEAN

Shikoku Island (Japan)

Ogasawara Islands (to Japan)

BRYDE'S WHALE IN SHIKOKU ISLAND, JAPAN

Trincomalee (Sri Lanka)

INDIAN OCEAN

AUSTRALASIA

Monkey Mia (Australia)

Hervey Bay (Australia)

Hermanus (South Africa)

Logan's Beach (Australia)

Kaikoura (New Zealand)

DUSKY DOLPHIN IN KAIKOURA, NEW ZEALAND

BOTTLENOSE DOLPHIN IN MONKEY MIA, AUSTRALIA

HOW TO WATCH CETACEANS

THERE ARE many ways of watching whales, dolphins, and porpoises around the world: from the air; from the shore; underwater; and from a host of different vessels, including yachts, motor cruisers, rubber inflatables, research boats, kayaks, and even huge ocean-going ships. The best commercial trips have experienced naturalists on board, who are skilled at finding the animals and provide interesting and informative commentaries, as well as skippers who are well versed in whale etiquette and behave responsibly. There are really just two golden rules for successful whale watching: the first and most important is to cause as little disturbance as possible; the second is to be patient.

PUTTING THE ANIMALS FIRST

Whale watching should be a strictly eyes-on, hands-off activity. Without due care and attention, propellers can cause serious injury, and the noise and movement of boats can cause the animals unnecessary stress. A simple code of conduct helps reduce the level of disturbance: never attempt a head-on approach; move slowly and do not go closer than 100 ft (30 m); do not separate or scatter a group of whales, dolphins, or porpoises; avoid sudden changes in speed or direction; do not stay longer than 15 minutes; avoid loud noises; and, when leaving, move off at a "no wake" speed until the boat is at least 985 ft (300 m) away. If a whale, dolphin, or porpoise approaches the boat (unless, of course, it is bow-riding), keep the engine in neutral and allow it to idle for approximately one minute before switching it off. From a safety point of view, ensure that you keep a sensible distance from particularly active whales.

KEEP YOUR DISTANCE
This is how not to watch cetaceans. They need plenty of space and should not feel trapped. Too many boats approaching too closely can be very stressful for the animals, and there is always the risk of propellers causing serious injury.

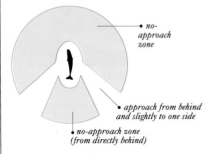

- no-approach zone
- approach from behind and slightly to one side
- no-approach zone (from directly behind)

HOW TO APPROACH A WHALE
A whale should be approached from a position slightly to the rear and to one side; as you get closer to the whale, continue along a parallel course.

USING A RANGEFINDER
Many places have guidelines or laws to protect cetaceans from irresponsible whale watchers. These vary according to location and species. Hawaii has some of the strictest regulations: to avoid coming too close, many skippers use a rangefinder to measure the distance between them and the animals.

EQUIPMENT

Several items of equipment are useful for whale watching. Binoculars (up to 10x) are invaluable for studying behavior and identifying species; a camera, preferably with motordrive and an 80–200 mm lens (or a similar zoom), helps to record observations; a notebook and pen and a stopwatch are useful for more detailed research; and a hydrophone can help find whales by sound, as well as adding a new dimension to the experience.

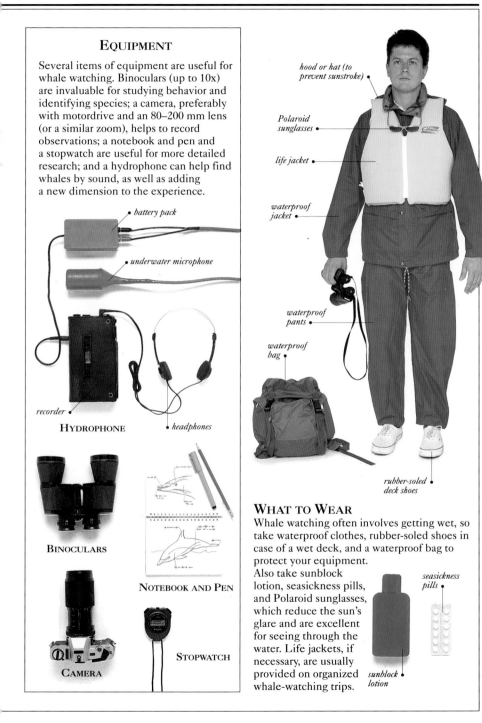

battery pack

underwater microphone

recorder

HYDROPHONE • headphones

BINOCULARS

NOTEBOOK AND PEN

STOPWATCH

CAMERA

hood or hat (to prevent sunstroke)

Polaroid sunglasses

life jacket

waterproof jacket

waterproof pants

waterproof bag

rubber-soled deck shoes

WHAT TO WEAR

Whale watching often involves getting wet, so take waterproof clothes, rubber-soled shoes in case of a wet deck, and a waterproof bag to protect your equipment. Also take sunblock lotion, seasickness pills, and Polaroid sunglasses, which reduce the sun's glare and are excellent for seeing through the water. Life jackets, if necessary, are usually provided on organized whale-watching trips.

seasickness pills

sunblock lotion

IDENTIFYING CETACEANS

IDENTIFYING WHALES, dolphins, and porpoises at sea is a challenge. Many species look alike; one individual is rarely identical to the next; they frequently spend long periods of time underwater; and many species tend to show little of themselves at the surface. Adverse conditions, such as heavy seas, high winds, driving rain, or even glaring sunshine, add to the difficulties. It is not surprising that even the world's experts are unable to identify every animal that they encounter – and many sightings have to be logged as "unidentified." However, with some background knowledge and a little practice, it is quite possible for anyone to recognize the more common and distinctive species and, eventually, many of the more unusual species as well.

IDENTIFICATION CHECKLIST

The key to successful identification is a simple process of elimination. This requires a mental checklist of the main features to look for in every whale, dolphin, or porpoise encountered at sea. One feature alone is rarely enough for a positive identification, so the golden rule is to gather information on as many of these features as possible before drawing any conclusions. There are 12 main points on this checklist:

1. Size
2. Unusual features (e.g. Narwhal's tusk)
3. Dorsal fin position, shape, and color
4. Body and head shape
5. Color and markings
6. Blow characteristics (in larger species only)
7. Fluke shape and markings
8. Surfacing behavior and dive sequence
9. Breaching and other activities
10. Number of animals observed
11. Main habitat (e.g. coastal, riverine)
12. Geographical location

Many items on the above list are explained in more detail on these two pages.

FLUKES

Observe whether the flukes are lifted in the air before the animal dives. When possible, make a note of the flukes' shape and any distinctive markings, and whether or not there is a notch between the trailing edges.

HOW LARGE IS THE GROUP?

Try to estimate the size of the group, as some species are more gregarious than others. As in this picture, there may be twice as many dolphins below the surface as there are above at any one time.

DORSAL FIN

Note whether the animal has a dorsal fin or hump. If it does, look carefully at its shape: does it have a broad or narrow base? Is it curved or upright? Is the tip pointed or rounded? Also, look at the size of the dorsal fin or hump in relation to the size of the body, its position on the animal's back, its color, and notice any distinctive markings it may have.

KILLER WHALE

HECTOR'S DOLPHIN

SOUTHERN/NORTHERN RIGHT WHALE

NARWHAL

SOWERBY'S BEAKED WHALE

OVERALL APPEARANCE

Try to estimate the animal's length and look at its overall shape: judge whether it has a robust or streamlined body and whether it has a distinctive beak. Note the main background color and any distinctive markings, such as body stripes or eye patches. Always bear in mind that colors at sea vary according to water clarity and light conditions. If you are facing the sun, remember that all animals will appear dark.

VAQUITA

INDO-PACIFIC
HUMP-BACKED DOLPHIN

FIN WHALE

FLIPPERS

It is not always possible to see the flippers, but they can be useful for identification of some species. Record their position on the animal's body, as well as their length, color, and shape. They can range from small and narrow to large and spatulate.

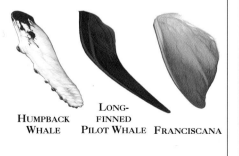

HUMPBACK
WHALE

LONG-
FINNED
PILOT WHALE   FRANCISCANA

IDENTIFICATION KEY

The Identification Key on the following pages (pp.30–37) is designed for quick reference at sea. The species are organized first by size, and then by the presence or absence of a prominent beak. Each entry includes both the common and scientific names, length, approximate geographical location, and a page reference to lead you to the detailed account of the species.

SIZE

It is difficult to estimate exact size at sea, unless a direct comparison can be made with the length of the boat or with another object in the water. For easy use, the key has three size categories, based on the species' typical size: up to 10 ft (3 m); 10–33 ft (3–10 m); and over 33 ft (10 m).

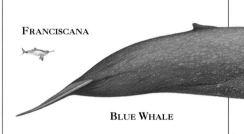

FRANCISCANA

BLUE WHALE

BEAK

The presence or absence of a prominent beak can be a very useful identification feature, especially in toothed species. Broadly speaking, river dolphins, beaked whales, and half the oceanic dolphins have prominent beaks, while porpoises, Belugas and Narwhals, blackfish, sperm whales, and the remaining oceanic dolphins do not. There is great variation in beak length from one species to another, but in this key the distinction is simply "with prominent beaks" and "without prominent beaks."

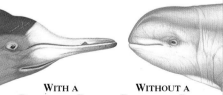

WITH A
PROMINENT BEAK

WITHOUT A
PROMINENT BEAK

IDENTIFICATION KEY

WHEN USING this Identification Key, bear in mind that the distinction between species "with" and "without" prominent beaks is purely to help with identification and is largely subjective; also, certain exceptions have been made to put similar species together. If the sexes look different, in most cases the illustration shows the more distinctive of the two (which is usually the male).

TYPICAL LENGTH UP TO 10 FT (3 M) WITH PROMINENT BEAKS

FRANCISCANA
Pontoporia blainvillei
S. hemisphere
4¼–5¾ ft (1.3–1.7 m) p.234

TUCUXI
Sotalia fluviatilis
N. and S. hemisphere
4¼–6 ft (1.3–1.8 m) p.172

SHORT-SNOUTED SPINNER DOLPHIN
Stenella clymene
N. hemisphere
5¼–6½ ft (1.7–2 m) p.180

COMMON DOLPHIN
Delphinus delphis
N. and S. hemisphere
5¼–8 ft (1.7–2.4 m) p.164

ATLANTIC SPOTTED DOLPHIN
Stenella frontalis
N. and S. hemisphere
5¾–7½ ft (1.7–2.3 m) p.186

PANTROPICAL SPOTTED DOLPHIN
Stenella attenuata
N. and S. hemisphere
5¾–8 ft (1.7–2.4 m) p.184

LONG-SNOUTED SPINNER DOLPHIN
Stenella longirostris
N. and S. hemisphere
4¼–7 ft (1.3–2.1 m) p.182

INDUS/GANGES RIVER DOLPHIN
*Platanista minor/
Platanista gangetica*
N. hemisphere
5–8¼ ft (1.5–2.5 m) p.230

BOTO
Inia geoffrensis
N. and S. hemisphere
6–8¼ ft (1.8–2.5 m) p.226

BAIJI
Lipotes vexillifer
N. hemisphere
4¾–8¼ ft (1.4–2.5 m) p.228

SOUTHERN RIGHTWHALE DOLPHIN
Lissodelphis peronii
S. hemisphere
6–9½ ft (1.8–2.9 m) p.170

STRIPED DOLPHIN
Stenella coeruleoalba
N. and S. hemisphere
6–8¼ ft (1.8–2.5 m) p.178

ROUGH-TOOTHED DOLPHIN
Steno bredanensis
N. and S. hemisphere
7–8½ ft (2.1–2.6 m) p.190

ATLANTIC HUMP-BACKED DOLPHIN
Sousa teuszii
N. and S. hemisphere
6½–8¼ ft (2–2.5 m) p.176

INDO-PACIFIC HUMP-BACKED DOLPHIN
Sousa chinensis
N. and S. hemisphere
6½–9¼ ft (2–2.8 m) p.174

NORTHERN RIGHTWHALE DOLPHIN
Lissodelphis borealis
N. hemisphere
6½–9¾ ft (2–3 m) p.168

BOTTLENOSE DOLPHIN
Tursiops truncatus
N. and S. hemisphere
6¼–12¾ ft (1.9–3.9 m) p.192

0   **39 in** (1 m)   **33 ft** (10 m)

TYPICAL LENGTH UP TO 10 FT (3 M) WITHOUT PROMINENT BEAKS

COMMERSON'S DOLPHIN
Cephalorhynchus commersonii
S. hemisphere
4¼–5¾ ft (1.3–1.7 m) p.198

HECTOR'S DOLPHIN
Cephalorhynchus hectori
S. hemisphere
4–5 ft (1.2–1.5 m) p.204

HEAVISIDE'S DOLPHIN
Cephalorhynchus heavisidii
S. hemisphere
5¼–5¾ ft (1.6–1.7 m) p.202

VAQUITA
Phocoena sinus
N. hemisphere
4–5 ft (1.2–1.5 m) p.244

FINLESS PORPOISE
Neophocaena phocaenoides
N. and S. hemisphere
4–6¼ ft (1.2–1.9 m) p.238

BLACK DOLPHIN
Cephalorhynchus eutropia
S. hemisphere
4–5¾ ft (1.2–1.7 m) p.200

HARBOR PORPOISE
Phocoena phocoena
N. hemisphere
4¾–6¼ ft (1.4–1.9 m) p.242

BURMEISTER'S PORPOISE
Phocoena spinipinnis
S. hemisphere
4¾–6½ ft (1.4–2 m) p.246

HOURGLASS DOLPHIN
Lagenorhynchus cruciger
S. hemisphere
c.5¼–6 ft (1.6–1.8 m) p.216

DUSKY DOLPHIN
Lagenorhynchus obscurus
S. hemisphere
5¼–7 ft (1.6–2.1 m) p.220

SPECTACLED PORPOISE
Australophocaena dioptrica
S. hemisphere
4¼–7¼ ft (1.3–2.2 m) p.240

PACIFIC WHITE-SIDED
DOLPHIN
Lagenorhynchus obliquidens
N. hemisphere
5¾–8 ft (1.7–2.4 m) p.218

PEALE'S DOLPHIN
Lagenorhynchus australis
S. hemisphere
c.6½–7¼ ft (2–2.2 m) p.214

DALL'S PORPOISE
Phocoenoides dalli
N. hemisphere
5¾–7¼ ft (1.7–2.2 m) p.248

ATLANTIC WHITE-SIDED
DOLPHIN
Lagenorhynchus acutus
N. hemisphere
6¼–8¼ ft (1.9–2.5 m) p.210

FRASER'S DOLPHIN
Lagenodelphis hosei
N. and S. hemisphere
6½–8½ ft (2–2.6 m) p.208

DWARF SPERM WHALE
Kogia simus
N. and S. hemisphere
7–9 ft (2.1–2.7 m) p.84

IRRAWADDY DOLPHIN
Orcaella brevirostris
N. and S. hemisphere
7–8½ ft (2.1–2.6 m) p.222

MELON-HEADED WHALE
Peponocephala electra
N. and S. hemisphere
7–9 ft (2.1–2.7 m) p.156

PYGMY KILLER WHALE
Feresa attenuata
N. and S. hemisphere
7–8½ ft (2.1–2.6 m) p.146

WHITE-BEAKED DOLPHIN
Lagenorhynchus albirostris
N. hemisphere
8¼–9¼ ft (2.5–2.8 m) p.212

RISSO'S DOLPHIN
Grampus griseus
N. and S. hemisphere
8½–12½ ft (2.6–3.8 m) p.206

PYGMY SPERM WHALE
Kogia breviceps
N. and S. hemisphere
9–11¼ ft (2.7–3.4 m) p.82

66 ft (20 m)

100 ft (30 m)

Typical Length 10–33 ft (3–10 m) With Prominent Beaks

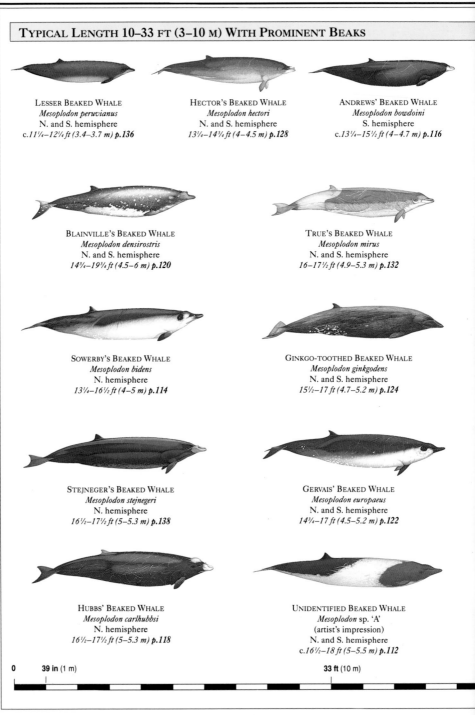

Lesser Beaked Whale
Mesoplodon peruvianus
N. and S. hemisphere
c.11¼–12¼ ft (3.4–3.7 m) **p.136**

Hector's Beaked Whale
Mesoplodon hectori
N. and S. hemisphere
13¼–14¾ ft (4–4.5 m) **p.128**

Andrews' Beaked Whale
Mesoplodon bowdoini
S. hemisphere
c.13¼–15½ ft (4–4.7 m) **p.116**

Blainville's Beaked Whale
Mesoplodon densirostris
N. and S. hemisphere
14¾–19¾ ft (4.5–6 m) **p.120**

True's Beaked Whale
Mesoplodon mirus
N. and S. hemisphere
16–17½ ft (4.9–5.3 m) **p.132**

Sowerby's Beaked Whale
Mesoplodon bidens
N. hemisphere
13¼–16½ ft (4–5 m) **p.114**

Ginkgo-toothed Beaked Whale
Mesoplodon ginkgodens
N. and S. hemisphere
15½–17 ft (4.7–5.2 m) **p.124**

Stejneger's Beaked Whale
Mesoplodon stejnegeri
N. hemisphere
16½–17½ ft (5–5.3 m) **p.138**

Gervais' Beaked Whale
Mesoplodon europaeus
N. and S. hemisphere
14¾–17 ft (4.5–5.2 m) **p.122**

Hubbs' Beaked Whale
Mesoplodon carlhubbsi
N. hemisphere
16½–17½ ft (5–5.3 m) **p.118**

Unidentified Beaked Whale
Mesoplodon sp. 'A'
(artist's impression)
N. and S. hemisphere
c.16½–18 ft (5–5.5 m) **p.112**

0   **39 in** (1 m)   **33 ft** (10 m)

TYPICAL LENGTH 10–33 FT (3–10 M) WITH PROMINENT BEAKS (continued)

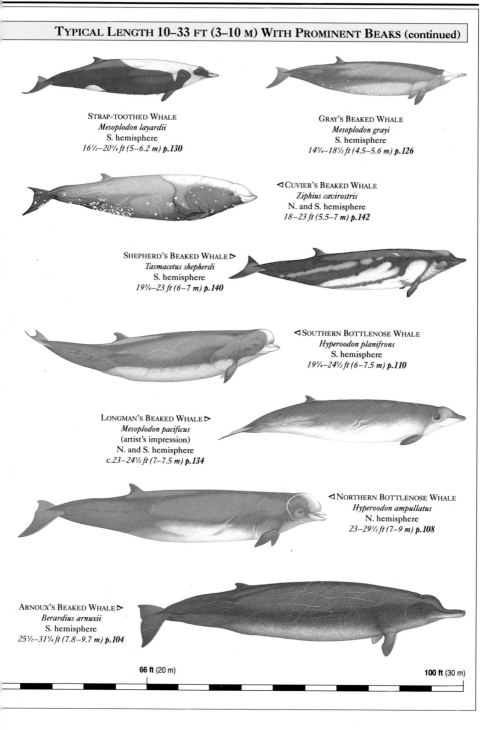

STRAP-TOOTHED WHALE
Mesoplodon layardii
S. hemisphere
16½–20¼ ft (5–6.2 m) **p.130**

GRAY'S BEAKED WHALE
Mesoplodon grayi
S. hemisphere
14¾–18½ ft (4.5–5.6 m) **p.126**

◁ CUVIER'S BEAKED WHALE
Ziphius cavirostris
N. and S. hemisphere
18–23 ft (5.5–7 m) **p.142**

SHEPHERD'S BEAKED WHALE ▷
Tasmacetus shepherdi
S. hemisphere
19¾–23 ft (6–7 m) **p.140**

◁ SOUTHERN BOTTLENOSE WHALE
Hyperoodon planifrons
S. hemisphere
19¾–24½ ft (6–7.5 m) **p.110**

LONGMAN'S BEAKED WHALE ▷
Mesoplodon pacificus
(artist's impression)
N. and S. hemisphere
c.23–24½ ft (7–7.5 m) **p.134**

◁ NORTHERN BOTTLENOSE WHALE
Hyperoodon ampullatus
N. hemisphere
23–29½ ft (7–9 m) **p.108**

ARNOUX'S BEAKED WHALE ▷
Berardius arnuxii
S. hemisphere
25½–31¼ ft (7.8–9.7 m) **p.104**

66 ft (20 m)

100 ft (30 m)

TYPICAL LENGTH 10–33 FT (3–10 M) WITHOUT PROMINENT BEAKS

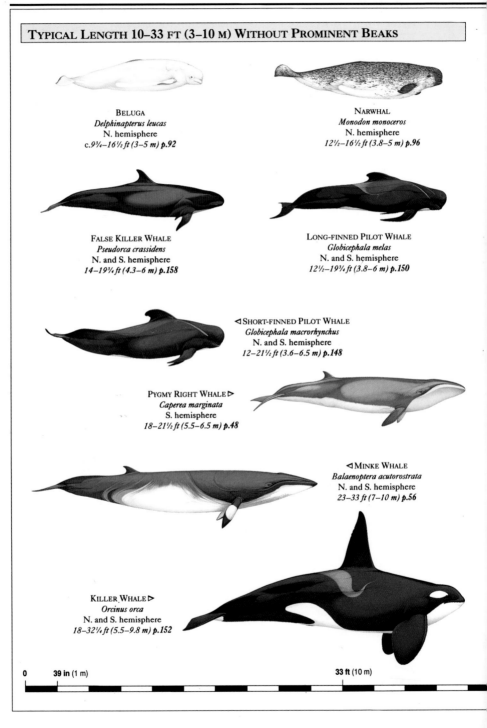

BELUGA
Delphinapterus leucas
N. hemisphere
c.9¾–16½ ft (3–5 m) p.92

NARWHAL
Monodon monoceros
N. hemisphere
12½–16½ ft (3.8–5 m) p.96

FALSE KILLER WHALE
Pseudorca crassidens
N. and S. hemisphere
14–19¾ ft (4.3–6 m) p.158

LONG-FINNED PILOT WHALE
Globicephala melas
N. and S. hemisphere
12½–19¾ ft (3.8–6 m) p.150

◁ SHORT-FINNED PILOT WHALE
Globicephala macrorhynchus
N. and S. hemisphere
12–21½ ft (3.6–6.5 m) p.148

PYGMY RIGHT WHALE ▷
Caperea marginata
S. hemisphere
18–21½ ft (5.5–6.5 m) p.48

◁ MINKE WHALE
Balaenoptera acutorostrata
N. and S. hemisphere
23–33 ft (7–10 m) p.56

KILLER WHALE ▷
Orcinus orca
N. and S. hemisphere
18–32¼ ft (5.5–9.8 m) p.152

0   **39 in** (1 m)   **33 ft** (10 m)

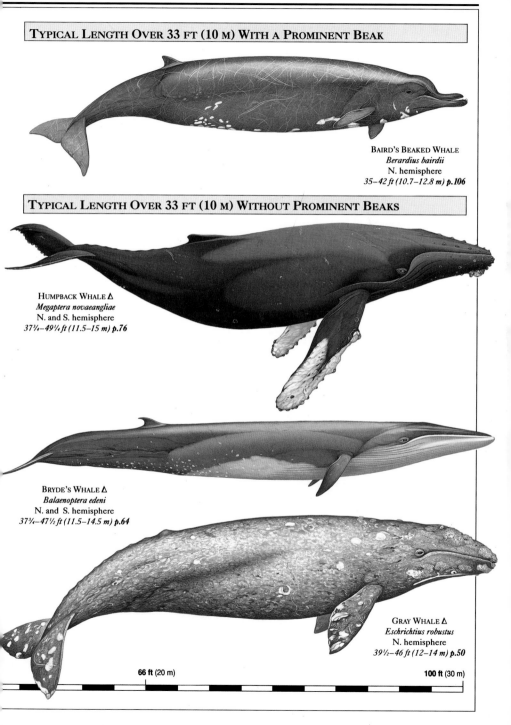

TYPICAL LENGTH OVER 33 FT (10 M) WITH A PROMINENT BEAK

BAIRD'S BEAKED WHALE
Berardius bairdii
N. hemisphere
35–42 ft (10.7–12.8 m) **p.106**

TYPICAL LENGTH OVER 33 FT (10 M) WITHOUT PROMINENT BEAKS

HUMPBACK WHALE Δ
Megaptera novaeangliae
N. and S. hemisphere
37¾–49¼ ft (11.5–15 m) **p.76**

BRYDE'S WHALE Δ
Balaenoptera edeni
N. and S. hemisphere
37¾–47½ ft (11.5–14.5 m) **p.64**

GRAY WHALE Δ
Eschrichtius robustus
N. hemisphere
39½–46 ft (12–14 m) **p.50**

66 ft (20 m)   **100 ft** (30 m)

TYPICAL LENGTH OVER 33 FT (10 M) WITHOUT PROMINENT BEAKS

△ SPERM WHALE
Physeter macrocephalus
N. and S. hemisphere
36–59 ft (11–18 m) **p.86**

△ SEI WHALE
Balaenoptera borealis
N. and S. hemisphere
39½–52½ ft (12–16 m) **p.60**

BLUE WHALE ▷
Balaenoptera musculus
N. and S. hemisphere
69–88½ ft (21–27 m) **p.68**

0   39 in (1 m)

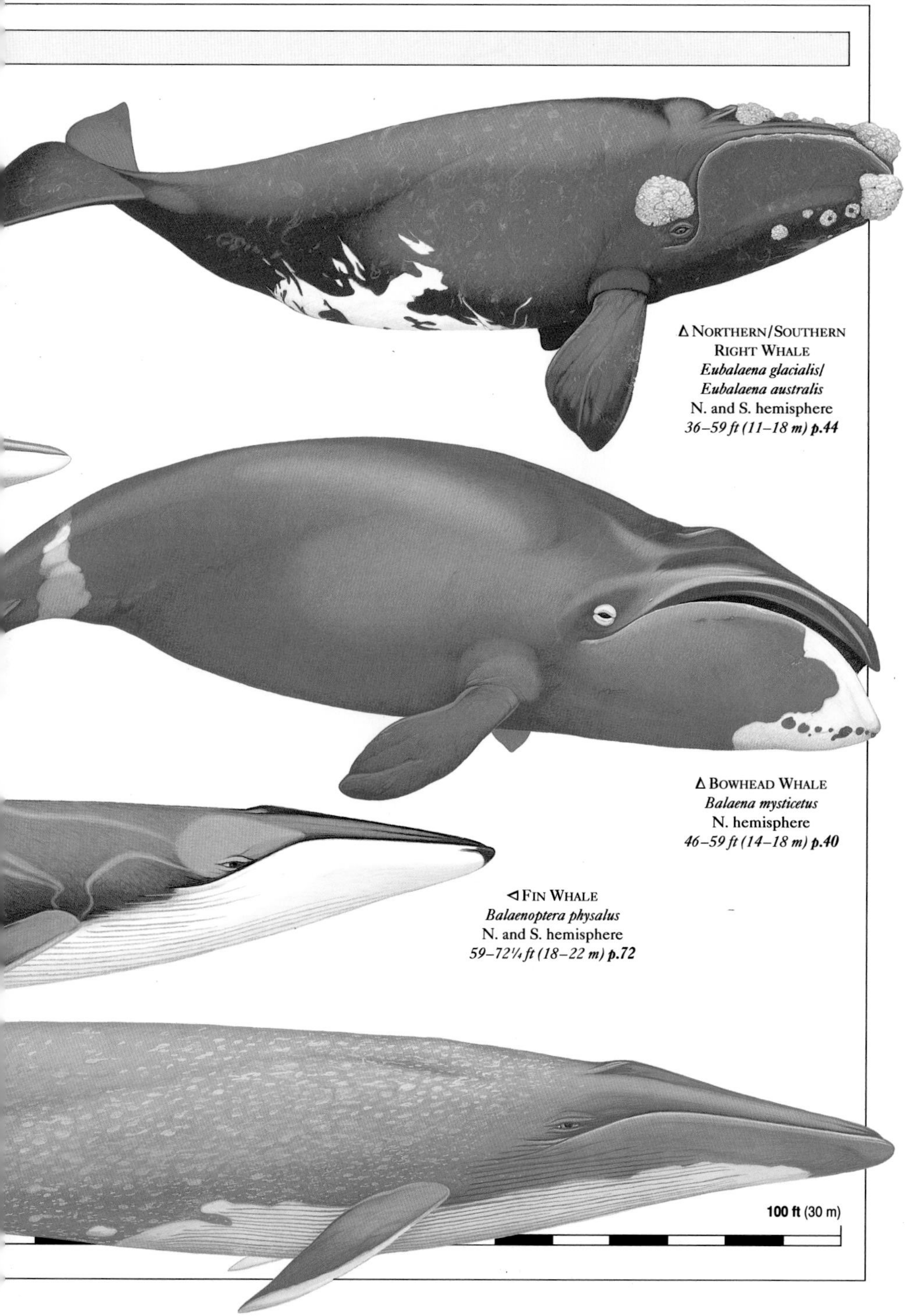

△ NORTHERN/SOUTHERN
RIGHT WHALE
*Eubalaena glacialis/
Eubalaena australis*
N. and S. hemisphere
36–59 ft (11–18 m) **p.44**

△ BOWHEAD WHALE
Balaena mysticetus
N. hemisphere
46–59 ft (14–18 m) **p.40**

◁ FIN WHALE
Balaenoptera physalus
N. and S. hemisphere
59–72¼ ft (18–22 m) **p.72**

100 ft (30 m)

RIGHT AND GRAY WHALES

T HREE VERY DIFFERENT families are included in this section: Balaenidae (Southern and Northern Right Whales and Bowhead Whale); Neobalaenidae (Pygmy Right Whale); and Eschrichtiidae (Gray Whale). The Southern and Pygmy Right Whales occur only in the southern hemisphere and the Bowhead Whale, Gray Whale, and Northern Right Whale are found only in the northern hemisphere. They all prefer temperate or polar waters. The Pygmy Right Whale has never been exploited by the whaling industry; however, the 4 larger species have suffered from tragic commercial overexploitation and all have, at some time, been close to extinction.

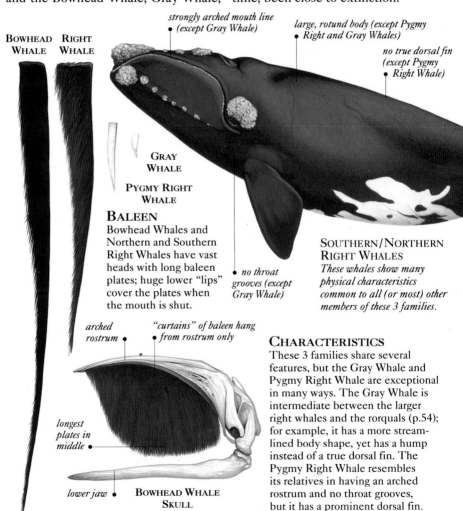

BOWHEAD WHALE RIGHT WHALE

strongly arched mouth line (except Gray Whale)

large, rotund body (except Pygmy Right and Gray Whales)

no true dorsal fin (except Pygmy Right Whale)

GRAY WHALE

PYGMY RIGHT WHALE

BALEEN
Bowhead Whales and Northern and Southern Right Whales have vast heads with long baleen plates; huge lower "lips" cover the plates when the mouth is shut.

no throat grooves (except Gray Whale)

SOUTHERN/NORTHERN RIGHT WHALES
These whales show many physical characteristics common to all (or most) other members of these 3 families.

arched rostrum

"curtains" of baleen hang from rostrum only

CHARACTERISTICS
These 3 families share several features, but the Gray Whale and Pygmy Right Whale are exceptional in many ways. The Gray Whale is intermediate between the larger right whales and the rorquals (p.54); for example, it has a more stream-lined body shape, yet has a hump instead of a true dorsal fin. The Pygmy Right Whale resembles its relatives in having an arched rostrum and no throat grooves, but it has a prominent dorsal fin.

longest plates in middle

lower jaw

BOWHEAD WHALE SKULL

CALLOSITIES
Southern and Northern Right Whales have growths, called callosities, above the eyes, by the blowholes, and on the chin, lower "lips," and rostrum.

broad flukes with
distinct notch in
middle

FLIPPERS
The flippers of the Bowhead Whale are considerably narrower and shorter (relative to body size) than those of Northern and Southern Right Whales.

| RIGHT | BOWHEAD |
| WHALES | WHALE |

PYGMY RIGHT WHALE
The Pygmy Right Whale has a distinctive way of swimming: it makes an undulatory movement that is caused by waves of motion passing from its head to its flukes (increasing in amplitude toward the rear of the body).

flippers extended
during slow swimming

flukes remain below
surface when diving

SPECIES IDENTIFICATION

PYGMY RIGHT WHALE
(p.48) *Smallest baleen whale, with a prominent dorsal fin.*

GRAY WHALE (p.50) *Arched head, and a low hump and "knuckles" instead of a dorsal fin.*

BOWHEAD WHALE (p.40) *White chin and a large head with no callosities, and no dorsal fin.*

NORTHERN RIGHT WHALE
(p.44) *Large head covered with callosities, and no dorsal fin.*

SOUTHERN RIGHT WHALE
(p.44) *Similar to the Northern Right Whale, but more common; lives only in the southern hemisphere.*

shows little (if any) of
dorsal fin or back at surface

S-shaped waves of
movement along entire body

PYGMY RIGHT WHALE DIVE SEQUENCE

Family BALAENIDAE	Species *Balaena mysticetus*	Habitat

BOWHEAD WHALE

Named for its distinctive bow-shaped skull, which is immense, the Bowhead Whale is a particularly heavy animal. No one has succeeded in weighing an entire specimen, in whole or in parts, but it is believed to be heavier for its body length than any other large whale. It is often found in association with Narwhals and Belugas, but is the only large whale living exclusively in the Arctic. Its blubber is up to 28 in (70 cm) thick, helping it to withstand the cold, and it can create its own breathing holes by breaking through ice up to 12 in (30 cm) thick. It was hunted to near extinction from a population of at least 50,000 in the mid-19th century. Its distinctive white chin, and the absence both of callosities and a dorsal fin should be sufficient for identification.

• **OTHER NAMES** Great Polar Whale, Arctic Whale, Arctic Right Whale, Greenland Right Whale, Greenland Whale.

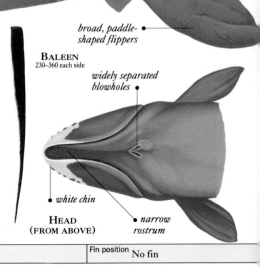

rounded back •

head approximately one-third of animal's entire length •

pronounced indentation behind blowhole •

huge, bowed mouth line •

"necklace" of black spots (variable) •

• *irregular white patch on chin (variable size)*

broad, paddle-shaped flippers •

BALEEN 230–360 each side

BEHAVIOR
Occasionally breaches, lobtails, flipper-slaps, and spyhops (usually alone). Young animals may play with objects in the water. It feeds at or below the surface and possibly along the sea floor; may move slowly at the surface with mouth open. Animals sometimes feed cooperatively. It is a slow swimmer. Typically spends 1 to 3 minutes at the surface, blowing 4 to 6 times. May dive to more than 655 ft (200 m); average dive time 4 to 20 minutes, but longer dives have been seen. Often surfaces again in the same place.

widely separated blowholes •

• *white chin*

HEAD
(FROM ABOVE)

• *narrow rostrum*

Group size 1–6 (1–14), loose groups of up to 60 (rare)	Fin position No fin

Status	Population	Threats
Rare	6,000–12,000	🚜 ☠ 🚢

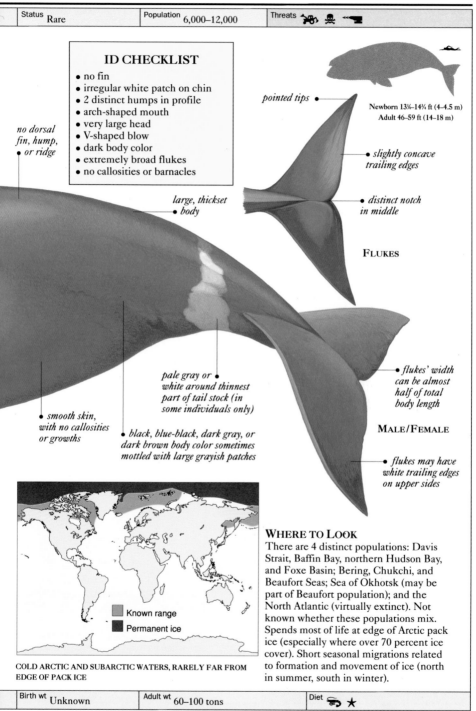

ID CHECKLIST

- no fin
- irregular white patch on chin
- 2 distinct humps in profile
- arch-shaped mouth
- very large head
- V-shaped blow
- dark body color
- extremely broad flukes
- no callosities or barnacles

no dorsal fin, hump, or ridge

pointed tips

Newborn 13¼–14¾ ft (4–4.5 m)
Adult 46–59 ft (14–18 m)

slightly concave trailing edges

large, thickset body

distinct notch in middle

FLUKES

flukes' width can be almost half of total body length

smooth skin, with no callosities or growths

pale gray or white around thinnest part of tail stock (in some individuals only)

black, blue-black, dark gray, or dark brown body color sometimes mottled with large grayish patches

MALE/FEMALE

flukes may have white trailing edges on upper sides

Known range

Permanent ice

COLD ARCTIC AND SUBARCTIC WATERS, RARELY FAR FROM EDGE OF PACK ICE

WHERE TO LOOK

There are 4 distinct populations: Davis Strait, Baffin Bay, northern Hudson Bay, and Foxe Basin; Bering, Chukchi, and Beaufort Seas; Sea of Okhotsk (may be part of Beaufort population); and the North Atlantic (virtually extinct). Not known whether these populations mix. Spends most of life at edge of Arctic pack ice (especially where over 70 percent ice cover). Short seasonal migrations related to formation and movement of ice (north in summer, south in winter).

Birth wt	Adult wt	Diet
Unknown	60–100 tons	🦐 ★

Family BALAENIDAE	Species *Balaena mysticetus*	Habitat 〰〰 (〰)

leaves water vertically •

half to three-quarters of body visible •

usually falls to one side •

SIDE VIEW

• *rounded back*

BREACHING
The Bowhead does not breach often, but when it does, it may do so repeatedly (the record is 64 times in 75 minutes). Typically, it comes out of the water vertically, the rear portion of the body usually remaining below the surface, and falls onto one side. Most breaching has been seen during spring migration.

calf usually more stubby and barrel-shaped than adult •

some calves born lighter in color (sometimes almost white) and darken with age •

• *pale bluish black body, sometimes appears light gray through water*

CALF

CALVES
Bowhead Whale calves are born from March to August, with a peak in May. The calf's size at birth can vary from 12 ft (3.6 m), which is a little below the average length, to as much as 17 ft (5.2 m). Its length doubles in the first year. Mature females probably give birth every 2 to 7 years. The mother–calf bond is extremely strong. This is one of the least known large whales, because its rarity and remote, harsh environment make research difficult.

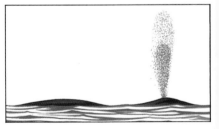

DIVE SEQUENCE
1. When surfacing, the whale shows 2 humps in profile; the blow projects directly upward in 2 columns (its V shape is less visible in wind).

Group size 1–6 (1–14), loose groups of up to 60 (rare)	Fin position No fin

Status Rare	Population 6,000–12,000	Threats

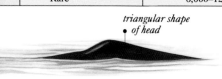

triangular shape of head

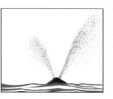

PROFILE

Most adult Bowhead Whales show 2 distinct humps in profile, reminiscent of the supposed shape of the Loch Ness Monster. The triangular hump in front is the head, the depression is the neck, and the rounded hump in the rear is the back, extending as far as the flukes. Note the smooth back, with no dorsal fin or ridge. Younger animals are rounder in profile and show a single rounded shape from the snout to the flukes.

BLOW
The widely separate blowholes produce a bushy, V-shaped blow rising to 23 ft (7 m).

FLUKING
The huge flukes, up to 23 ft (7 m) across, are usually thrown into the air before a deep dive.

most plates dark gray or black (fringes may be lighter) •

• *pale blue flukes*

longest plates in middle of jaw •

• *trailing edges may be slightly convex*

BALEEN

The Bowhead Whale has the longest baleen plates of any whale: many have been recorded over 9¾ ft (3 m) long and there is a disputed claim of one reaching 19 ft (5.8 m). They often show a green iridescence in bright sunlight. As with the Northern and Southern Right Whales, there are no baleen plates at the front of the mouth.

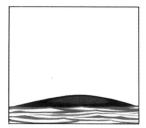

2. The head disappears below the surface, and the smooth, rounded back is arched in preparation for a long dive.

3. The broad flukes are typically lifted high into the air; the tail often tips to the right side as the flukes slip into the water.

4. The tail drops below the surface, often with the tip of the left fluke disappearing last. The dive may be more than 655 ft (200 m) deep.

Birth wt Unknown	Adult wt 60–100 tons	Diet

Family BALAENIDAE	Species *Eubalaena australis* (Southern)	Habitat 〰️ (〰️)

SOUTHERN AND NORTHERN RIGHT WHALES

It is uncertain whether there are 1, 2, or 3 species of the genus *Eubalaena*. Most authorities recognize 2, but animals in the North Pacific have been proposed as a third *(Eubalaena japonica)*. There are minor cranial differences between the northern and southern hemisphere animals, and some authorities suggest that southern animals have more callosities on top of their lower "lip" and fewer on top of the head. Technical data for the Southern Right Whale appears in the colored bands on these pages; for the Northern Right Whale, see pp.46–47.

• **OTHER NAMES** Black Right Whale, Right Whale (both species); Biscayan Right Whale (Northern Right Whale).

largest callosity on tip of rostrum • (known as "bonnet")

strongly arched mouth line •

• callosities (on head only)

• dark chin

"finger" bones can be traced by prominent ridges •

• large, spatulate flippers

BEHAVIOR

Slow, lumbering swimmers, but surprisingly acrobatic. May be seen waving flippers above the surface, breaching, flipper-slapping, and lobtailing. Southern Right Whales will also "head-stand," waving flukes high in the air for up to 2 minutes. Both species sometimes swim near the surface with mouths agape, showing baleen. Sometimes approachable. Playful and inquisitive: will poke, bump, and push objects in the water. Members of small groups may take turns to surface, with only a single animal visible at a time. Bellowing sounds and moans commonly heard in breeding areas, mostly at night. Rarely strand.

BALEEN
205–270 each side

narrow rostrum •

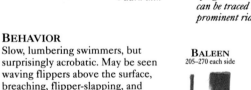

HEAD (FROM ABOVE)

paired • blowholes widely separated

Group size 2–3 (1–12), more at feeding grounds	Fin position No fin

Status		Population		Threats
Rare		3,000–5,000		🪢 🌊 🐟

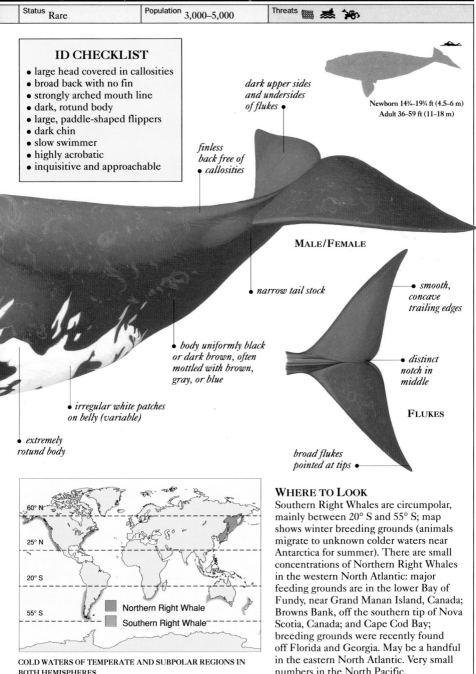

ID CHECKLIST

- large head covered in callosities
- broad back with no fin
- strongly arched mouth line
- dark, rotund body
- large, paddle-shaped flippers
- dark chin
- slow swimmer
- highly acrobatic
- inquisitive and approachable

dark upper sides and undersides of flukes •

Newborn 14¾–19¾ ft (4.5–6 m)
Adult 36–59 ft (11–18 m)

finless back free of • callosities

MALE/FEMALE

• narrow tail stock

• smooth, concave trailing edges

• body uniformly black or dark brown, often mottled with brown, gray, or blue

• distinct notch in middle

FLUKES

• irregular white patches on belly (variable)

• extremely rotund body

broad flukes pointed at tips •

WHERE TO LOOK

Southern Right Whales are circumpolar, mainly between 20° S and 55° S; map shows winter breeding grounds (animals migrate to unknown colder waters near Antarctica for summer). There are small concentrations of Northern Right Whales in the western North Atlantic: major feeding grounds are in the lower Bay of Fundy, near Grand Manan Island, Canada; Browns Bank, off the southern tip of Nova Scotia, Canada; and Cape Cod Bay; breeding grounds were recently found off Florida and Georgia. May be a handful in the eastern North Atlantic. Very small numbers in the North Pacific.

60° N
25° N
20° S
55° S

■ Northern Right Whale
■ Southern Right Whale

COLD WATERS OF TEMPERATE AND SUBPOLAR REGIONS IN BOTH HEMISPHERES

Birth wt		Adult wt		Diet
1 ton		30–80 tons		🦐

Family BALAENIDAE	Species *Eubalaena glacialis* (Northern)	Habitat 〰️ (〰️)

ENDANGERED SPECIES

Right whales were originally named by whalers because they were the "right" whales to catch: they were easy to approach, lived close to shore, floated when they were dead, and provided large quantities of valuable oil, meat, and whalebone. Both species came very close to extinction, but they have been protected since 1937. Only the Southern Right Whale is showing significant signs of recovery, with an annual increase of approximately 7 percent in recent years. The Northern Right Whale is probably closer to extinction than any other large whale and may never recover; it may already be doomed to extinction in the eastern North Atlantic, where it was formerly abundant between the Azores and Spitsbergen. Unfortunately, both species are slow breeders: females have their first calves at 5 to 10 years and give birth every 3 to 4 years.

calves have few or no callosities

body less rotund, even slender, in younger animals

CALF

some calves born lighter in color and darken with age

parasitic cyamid crustacean

flukes raised at right angles to wind

WHALE LICE

Whale lice, or cyamid crustaceans, live on the callosities and make them appear white, pink, yellow, or orange. Barnacles and parasitic worms may also live on these growths.

SAILING

Southern Right Whales sometimes raise their flukes at right angles to the wind and use them as sails, allowing themselves to be blown along through the water. This appears to be a form of play, because they will often swim back to the starting point and do it again.

DIVE SEQUENCE

1. The head is held high up out of the water, showing callosities. The left side of the V-shaped blow is taller than the right.

Group size 1–3 (1–12), more at feeding grounds	Fin position No fin

Status Endangered	Population 300–600	Threats 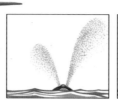

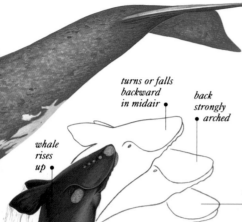

turns or falls backward in midair

back strongly arched

whale rises up

BLOW
The wide, V-shaped blow is possibly 16½ ft (5 m) high and may appear as one jet from the side or in the wind.

FLUKING
The flukes are often thrown high into the air before a deep dive. Note the smooth trailing edges and distinct notch.

lands with a distinctive wall of spray on each side

BREACHING
Right whales often breach, sometimes up to 10 times or more in a row. A huge wall of spray rises on each side as the whale hits the water. The splash can be heard from ¾ mile (1 km) away or more.

dense but fine bristles

exceptionally long, narrow baleen plates

BALEEN
The baleen varies from dark brown to dark gray or black, though it can appear yellowish underwater and is lighter in younger animals. The long, narrow rostrum is designed for the suspension of the plates, and the bowed lower jaw is for closing the mouth.

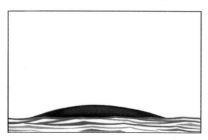  

2. The head disappears below the surface. All that is visible is the smooth, broad, finless back, which is very distinctive. It is free of barnacles and callosities.

3. The flukes are usually raised when diving, but the whale may "false-fluke" and not lift them out of the water.

4. The whale drops below the surface vertically; the dive may last up to an hour but is usually much shorter.

Birth wt 1 ton	Adult wt 30–80 tons	Diet

Family NEOBALAENIDAE	Species *Caperea marginata*	Habitat ▶▧ ▧

PYGMY RIGHT WHALE

The Pygmy Right Whale is the smallest and least known of all the baleen whales. It is seldom seen at sea, and confirmed sightings are very rare. It is difficult to distinguish from a Minke Whale (p.56), but there are visible differences: the Pygmy Right Whale has a strongly arched jawline, and many Minkes have distinctive white flipper bands. The other, considerably larger, right whales share the arched mouth. However, unlike these, the Pygmy Right Whale has a dorsal fin; the flipper shape is also very different, and it has a more streamlined body and a smaller head in relation to its overall size than the larger species. With such limited information available, it is impossible to estimate the Pygmy Right Whale's abundance, but it may be more common than the few sightings so far suggest.
• **OTHER NAMES** None.

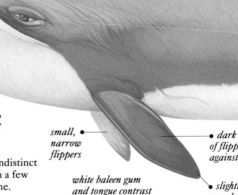

small ridge runs from near blowhole to tip of snout •

blowholes in slight depression •

head about one-quarter of body length •

dark gray or blue-gray upper side •

light lower jaw, dark rostrum •

jaw may become more arched with age •

rounded lump at base of throat •

small, narrow flippers •

• *dark upper sides of flippers stand out against paler body*

white baleen gum and tongue contrast with dark head •

• *slightly rounded tips*

BEHAVIOR
Inconspicuous at sea, with a small, indistinct blow. Typically spends no more than a few seconds at the surface at any one time. Breaching and lobtailing have not been observed. Surfaces like a Minke Whale, "throwing" its snout out of the water; but unlike the Minke, its dorsal fin and back sometimes remain hidden from view. A flash of white may be seen (the lower jaw or baleen gum) as the snout breaks the surface. The flukes are never lifted clear of the water. Generally swims slowly, in an unusual, undulating style, with waves of movement along the whole body, but is capable of rapid acceleration. Limited evidence suggests that longer dives last between 40 seconds and 4 minutes. Has been observed associating with Long-finned Pilot Whales, Sei Whales, and possibly Minke Whales.

MOUTH

BALEEN
213–230 each side

• *ivory-colored baleen*

Group size 1–2 (1–8)	Fin position Far behind center

Status Unknown	Population Unknown	Threats

ID CHECKLIST

- strongly arched mouth
- pronounced fin
- gray upper side, paler underside
- white baleen gum
- dark upper sides of flippers
- no callosities on head
- small size
- undemonstrative behavior
- slow swimmer

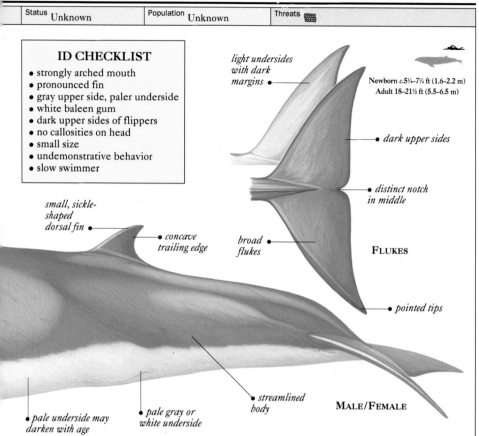

light undersides with dark margins •

Newborn *c*.5¼–7¼ ft (1.6–2.2 m)
Adult 18–21½ ft (5.5–6.5 m)

• *dark upper sides*

• *distinct notch in middle*

small, sickle-shaped dorsal fin •

• *concave trailing edge*

broad flukes •

FLUKES

• *pointed tips*

small, sickle-shaped dorsal fin •

• *pale underside may darken with age*

• *pale gray or white underside*

• *streamlined body*

MALE/FEMALE

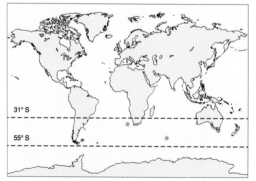

31° S

55° S

TEMPERATE WATERS OF THE SOUTHERN HEMISPHERE, BOTH INSHORE AND OFFSHORE

WHERE TO LOOK

Mostly known from widely dispersed strandings, especially in New Zealand, southern Australia, and South Africa, but little information is available. Most records within 31° S and 52° S, although there is one from Tierra del Fuego, southern South America, at 55° S. Real limit seems to be the surface water temperature: rarely found outside 41°–68° F (5°–20° C). Juveniles may migrate into inshore waters during spring and summer. At least some populations are year-round residents, for example in Tasmania. Most observations have been in sheltered, shallow bays, but several individuals have been encountered in the open sea.

Birth wt Unknown	Adult wt 3–3.5 tons	Diet

Family ESCHRICHTIIDAE	Species *Eschrichtius robustus*	Habitat 〰

GRAY WHALE

One of the most observed of all whales, the Gray Whale is well known for the 12,000-mile (19,500-km) round trip between its southern breeding grounds in Baja California, Mexico, and its northern feeding grounds in the Bering, Chukchi, and western Beaufort seas. This is one of the longest known migrations of any mammal. Whaling has taken a heavy toll: the North Atlantic population was extinct by the 17th or 18th century; the Korean population, on the western side of the North Pacific, was feared extinct, but a small number may possibly survive; and the Californian population, on the eastern side of the Pacific, was reduced to a few hundred or thousand by the early 1900s. The species was protected in 1946 (though a noncommercial quota is still set every year to be taken by Siberian and Alaskan native peoples), and the Californian population has since made a remarkable recovery. It exhibits many features intermediate between the right whales and the rorquals (see p.38).

• **OTHER NAMES** California Gray Whale, Devilfish, Mussel-digger, Scrag Whale.

head arches between blowhole and snout

blowholes open in shallow depression on top of head

long, slightly arched or straight mouth line

long, slender head, small in relation to body size

small, paddle-shaped flippers

pointed tips

BALEEN 140–180 each side

BEHAVIOR

One of the more active of all large whales: spy-hopping, lobtailing, and breaching commonly observed. Enjoys surf-riding and frequently found in surf (especially at Baja California) in very shallow water. May also lie on side at the surface and wave a flipper in the air. When migrating, typically blows 3 to 6 times (at 15- to 30-second intervals) before diving for 3 to 5 minutes. Cruising speed 2 to 5 knots. Dive sequence much more variable at feeding and breeding grounds: often changes course and may stay under for up to 18 minutes. Dives for food in water up to 395 ft (120 m) deep, but prefers much shallower water. Primarily a bottom-feeder. When feeding, typically followed by clouds of mud stirred up from sea floor or coming from mouth during filtration. May associate with several species of dolphin and Dall's Porpoise.

typically 2, but sometimes 3–7, V-shaped or parallel grooves on throat

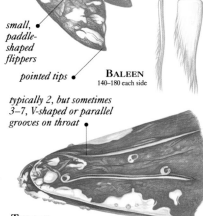

THROAT

Group size 1–3 (1–18), larger congregations in some areas	Hump position Far behind center

Status	Population	Threats
Locally common	c.15,000–25,000	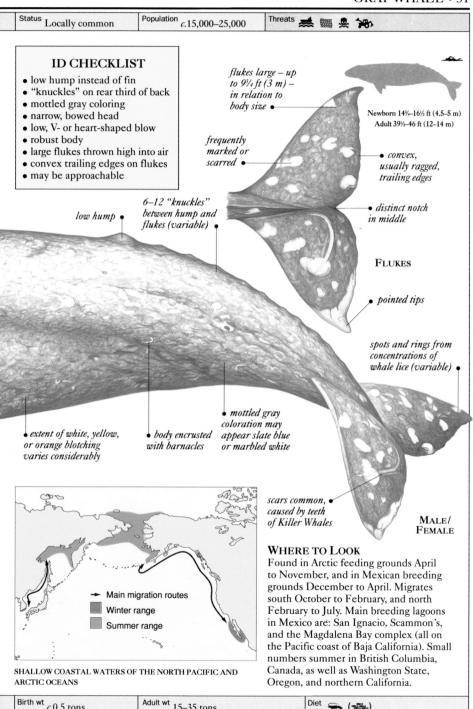

ID CHECKLIST

- low hump instead of fin
- "knuckles" on rear third of back
- mottled gray coloring
- narrow, bowed head
- low, V- or heart-shaped blow
- robust body
- large flukes thrown high into air
- convex trailing edges on flukes
- may be approachable

flukes large – up to 9¾ ft (3 m) – in relation to body size

Newborn 14¾–16½ ft (4.5–5 m)
Adult 39½–46 ft (12–14 m)

frequently marked or scarred

convex, usually ragged, trailing edges

distinct notch in middle

FLUKES

low hump

6–12 "knuckles" between hump and flukes (variable)

pointed tips

spots and rings from concentrations of whale lice (variable)

mottled gray coloration may appear slate blue or marbled white

extent of white, yellow, or orange blotching varies considerably

body encrusted with barnacles

scars common, caused by teeth of Killer Whales

MALE/ FEMALE

→ Main migration routes
■ Winter range
■ Summer range

SHALLOW COASTAL WATERS OF THE NORTH PACIFIC AND ARCTIC OCEANS

WHERE TO LOOK

Found in Arctic feeding grounds April to November, and in Mexican breeding grounds December to April. Migrates south October to February, and north February to July. Main breeding lagoons in Mexico are: San Ignacio, Scammon's, and the Magdalena Bay complex (all on the Pacific coast of Baja California). Small numbers summer in British Columbia, Canada, as well as Washington State, Oregon, and northern California.

Birth wt	Adult wt	Diet
c.0.5 tons	15–35 tons	

Family ESCHRICHTIIDAE	Species *Eschrichtius robustus*	Habitat

CALVES

Pregnant females give birth to single calves just before or soon after arriving at the breeding lagoons; in Mexico, most births occur between January 5 and February 15 (with a peak around January 27). Mothers and young calves usually remain in inner stretches of the lagoons, away from males and single females. They were nicknamed "Devilfish" by Yankee whalers, because the females are extremely protective of their young and frequently chased or attacked the whalers; today they are better known for being friendly toward people.

calves are born with wrinkles, which rapidly disappear after birth •

CALF

• yellowish, coarse baleen with long, thick bristles

HEAD

• right side often scarred from bottom-feeding

each louse up to 1 in • (2.5 cm) long

PARASITES

Gray Whales are more heavily infested with external parasites than any other whale and carry both lice and barnacles.

lice infect • barnacle clusters and folds of skin

FEEDING

The Gray Whale is unique among whales in being a bottom-feeder. It rolls onto its right side (though some individuals are "left-handed" and roll to the left) and sucks up sediment containing benthic amphipods from the sea floor; the water and silt are sieved out through the baleen with the help of the tongue. Consequently, the plates on the right side are usually shorter and more worn than those on the left. The right side of the whale's head is often scarred for the same reason.

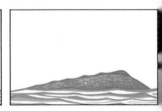

DIVE SEQUENCE
1. As the whale blows, its head appears to slope downward from the blowholes, giving the appearance of a shallow triangle.

2. After the final blow, the "knuckles" along the whale's back appear and the body takes on a deeper triangular shape.

Group size 1–3 (1–18), larger congregations in some areas	Hump position Far behind center

Status Locally common	Population *c*.15,000–25,000	Threats 🌊 ▦ ☠ 🚜

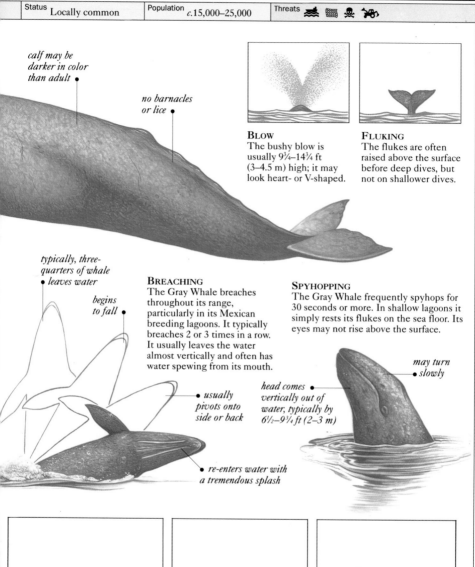

calf may be darker in color than adult •

no barnacles or lice •

BLOW
The bushy blow is usually 9¾–14¾ ft (3–4.5 m) high; it may look heart- or V-shaped.

FLUKING
The flukes are often raised above the surface before deep dives, but not on shallow dives.

typically, three-quarters of whale • *leaves water*

begins to fall •

BREACHING
The Gray Whale breaches throughout its range, particularly in its Mexican breeding lagoons. It typically breaches 2 or 3 times in a row. It usually leaves the water almost vertically and often has water spewing from its mouth.

SPYHOPPING
The Gray Whale frequently spyhops for 30 seconds or more. In shallow lagoons it simply rests its flukes on the sea floor. Its eyes may not rise above the surface.

may turn • *slowly*

• *usually pivots onto side or back*

head comes • *vertically out of water, typically by 6½–9¾ ft (2–3 m)*

• *re-enters water with a tremendous splash*

3. The main hump disappears below the surface and the back becomes more rounded. The "knuckles" are still visible above the surface.

4. The tail stock and flukes emerge above the surface and rise into the air in preparation for a deep dive.

5. As the flukes are raised higher into the air, they help thrust the whale's bulk into a more steeply angled dive.

Birth wt *c*.0.5 tons	Adult wt 15–35 tons	Diet 🐟 (🦐)

RORQUAL WHALES

THE RORQUALS are all large whales. The longest, the Blue Whale, is typically around 82 ft (25 m) long and is one of the biggest animals ever to have lived on earth; even the smallest, the Minke Whale, can grow as long as 33 ft (10 m). Females are slightly larger than males, in all 6 members of the family, and southern-hemisphere animals tend to be larger than those in the northern hemisphere. With the exception of the Bryde's Whale, the larger rorquals generally migrate long distances between warm-water winter breeding grounds and cold-water summer feeding grounds. The Humpback Whale is the odd member of the family, having a stockier body and longer flippers, and is placed in its own genus. The rorqual family has suffered from intensive exploitation by the whaling industry, and many populations are severely depleted or have disappeared.

single ridge along top of head (Bryde's Whale has 3) •

broad, flat head with pointed snout •

double blowhole •

CHARACTERISTICS

The word "rorqual" comes from the Norwegian word *rorhval*, which means "furrow"; the name refers to the many folds of skin, or throat grooves, that extend from underneath the lower jaw to behind the flippers in all members of this family. No other cetaceans have so many or such highly developed grooves, although the Gray Whale has up to 4 simple grooves and beaked whales have V-shaped grooves under the chin. The grooves allow a tremendous expansion of the mouth cavity but are rarely seen when the whales are not feeding.

12–100 longitudinal throat grooves, or pleats, depending on individual and species •

long, slender body (except Humpback Whale) •

FEEDING

Rorquals have many different feeding techniques, but they all work on the same principle: they take in tons of water in a mouthful and, with their baleen, filter out the fish or krill. They have up to 100 grooves, or pleats, on their throats, which expand and contract like a concertina to hold the vast quantities of food-laden water. This efficient system enables some of the largest animals in the world to feed on some of the smallest.

RORQUAL FEEDING

1. Whale seeks out a good feeding area, where water is teeming with fish or krill.

2. It swims along with its mouth open, taking in huge quantities of seawater.

3. Sheer volume of water begins to distend its throat as pleats pull apart.

4. Pleats contract as mouth closes; water is forced out, and prey is caught inside baleen.

MINKE WHALE (p.56) *Smallest and most common rorqual; has a pointed snout and an indistinct blow; many have a white flipper band.*

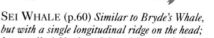

BRYDE'S WHALE

The 3 ridges on the head of the Bryde's Whale are unique to this species and, at close range, are an unmistakable identification feature.

BRYDE'S WHALE (p.64) *Easily recognizable at close range by 3 parallel longitudinal ridges on the head; skin is often mottled with circular scars.*

• *dorsal fin set far back on body*

HUMPBACK WHALE (p.76)
Unmistakable whale with a knobbly head and the largest flippers of any cetacean; flukes are usually lifted high into the air before a long dive.

MINKE WHALE

The Minke Whale is the smallest of the rorquals but illustrates many of the physical characteristics common to most members of the family. The most significant difference between species is size.

SEI WHALE (p.60) *Similar to Bryde's Whale, but with a single longitudinal ridge on the head; has a tall, sickle-shaped dorsal fin; poorly known.*

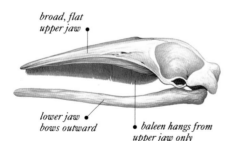

broad, flat upper jaw •

lower jaw • *bows outward*   • *baleen hangs from upper jaw only*

FIN WHALE (p.72) *Exceptionally long, streamlined body with a backward-leaning dorsal fin and asymmetrical head pigmentation; body does not have any mottling.*

SKULL

The rorquals have baleen plates, instead of teeth, hanging from their upper jaws. Rorqual baleen is relatively broader and shorter than that of other baleen whales and, consequently, the upper jaw is not as strongly arched.

BLUE WHALE (p.68) *Enormous whale – it is almost as long as a Boeing 737 and is the largest of all cetaceans; has mottled blue-gray coloring and a tiny, stubby dorsal fin.*

Family BALAENOPTERIDAE	Species *Balaenoptera acutorostrata*	Habitat

MINKE WHALE

The Minke Whale is the smallest and most abundant of the rorquals. It is highly variable in appearance, and some authorities recognize 3 or even 4 subspecies. Some animals are inquisitive and approach quite closely, but in most cases it is unusual to get a clear view. At a distance, it may be confused with the Sei Whale (p.60), Bryde's Whale (p.64), Fin Whale (p.72), or Northern Bottlenose Whale (p.108), but the dive sequence of the Minke is quite distinctive. The head shape and relatively unscarred skin should be sufficient to tell a Minke from most beaked whales (p.100), and its relatively straight mouth line should help distinguish it from a Pygmy Right Whale (p.48). Animals in the northern hemisphere have a white band on the flippers, but this is absent on many southern-hemisphere animals. The Minke is now the only baleen whale being hunted commercially.

• **OTHER NAMES** Pikehead, Little Piked Whale, Pike Whale, Little Finner, Sharp-headed Finner, Lesser Finback, Lesser Rorqual.

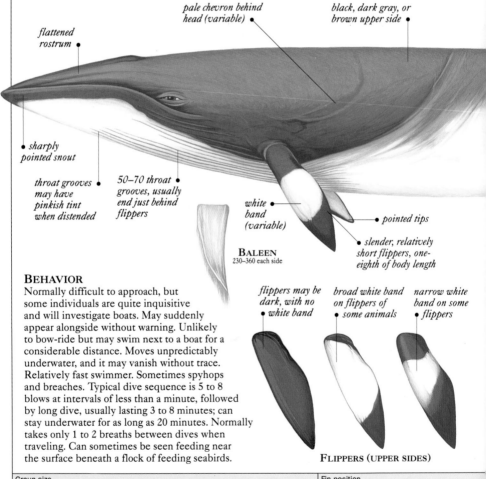

pale chevron behind head (variable) •

black, dark gray, or brown upper side •

flattened rostrum •

• sharply pointed snout

throat grooves may have pinkish tint when distended

50–70 throat grooves, usually end just behind flippers

white • band (variable)

• pointed tips

BALEEN
230–360 each side

• slender, relatively short flippers, one-eighth of body length

BEHAVIOR

Normally difficult to approach, but some individuals are quite inquisitive and will investigate boats. May suddenly appear alongside without warning. Unlikely to bow-ride but may swim next to a boat for a considerable distance. Moves unpredictably underwater, and it may vanish without trace. Relatively fast swimmer. Sometimes spyhops and breaches. Typical dive sequence is 5 to 8 blows at intervals of less than a minute, followed by long dive, usually lasting 3 to 8 minutes; can stay underwater for as long as 20 minutes. Normally takes only 1 to 2 breaths between dives when traveling. Can sometimes be seen feeding near the surface beneath a flock of feeding seabirds.

flippers may be dark, with no • white band

broad white band on flippers of • some animals

narrow white band on some • flippers

FLIPPERS (UPPER SIDES)

Group size 1 (1–3), occasionally up to 100 or more at good feeding areas	Fin position Far behind center

Status	Population	Threats
Common	*c.*500,000–1 million	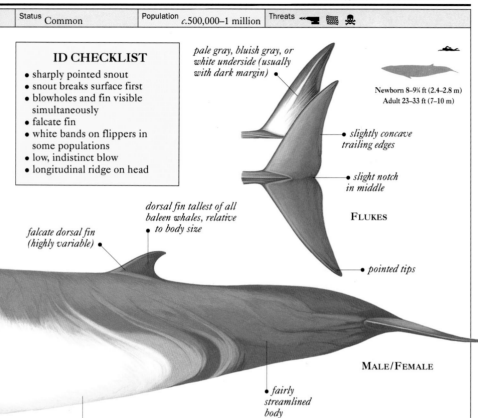

ID CHECKLIST

- sharply pointed snout
- snout breaks surface first
- blowholes and fin visible simultaneously
- falcate fin
- white bands on flippers in some populations
- low, indistinct blow
- longitudinal ridge on head

pale gray, bluish gray, or white underside (usually with dark margin)

Newborn 8–9¼ ft (2.4–2.8 m)
Adult 23–33 ft (7–10 m)

• slightly concave trailing edges

• slight notch in middle

FLUKES

dorsal fin tallest of all baleen whales, relative to body size

falcate dorsal fin (highly variable) •

• pointed tips

MALE/FEMALE

• fairly streamlined body

• white, pale gray, or pale brown underside

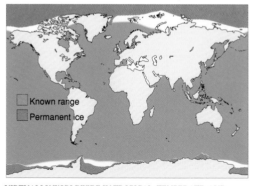

VIRTUALLY WORLDWIDE IN TROPICAL, TEMPERATE, AND POLAR WATERS OF BOTH HEMISPHERES

WHERE TO LOOK

Found virtually worldwide, but probably not continuous distribution. Generally less common in the tropics than in cooler waters. Three geographically isolated populations are recognized: in the North Pacific, North Atlantic, and southern hemisphere. Usually more concentrated in higher latitudes during summer and lower latitudes during winter, but migrations vary from year to year. Some populations appear to be resident year-round, and recent evidence suggests that individuals may have exclusive home ranges in some areas. Often enters estuaries, bays, and inlets, and during summer, may feed around headlands and small islands. Sometimes gets trapped inside small pockets of open water within pack ice.

☐ Known range
☐ Permanent ice

Birth wt	Adult wt	Diet
*c.*770 lb (350 kg)	5–10 tons	

Family BALAENOPTERIDAE	Species *Balaenoptera acutorostrata*	Habitat

BREACHING

Minke Whales have been observed breaching on many occasions, though not as often as some of their larger relatives. They usually leave the water dorsal side up, at an angle of about 45°, and re-enter without twisting or turning their bodies. Most of the body may leave the water in the initial surge (at least as far as the tail stock) and the entire dorsal fin is often visible. The back can be arched in midair or held fairly straight, as shown here. Sometimes they land on their stomachs with a tremendous splash but, equally, they may re-enter head first in a clean, dolphinlike dive. A breach is often repeated 2 or 3 times in a row, although longer sequences have been observed.

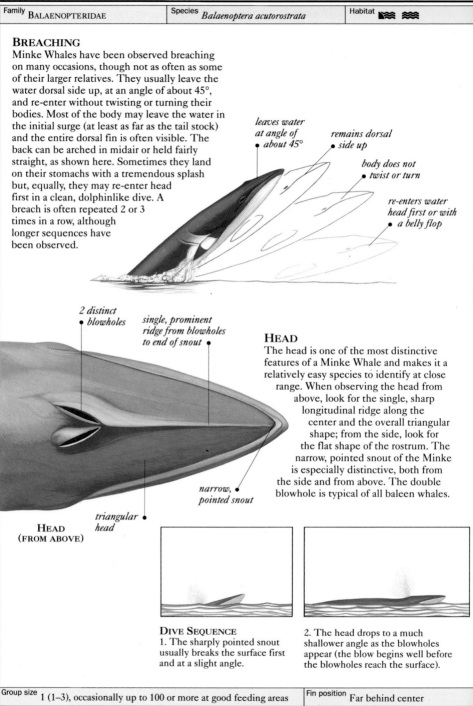

leaves water at angle of about 45°

remains dorsal side up

body does not twist or turn

re-enters water head first or with a belly flop

2 distinct blowholes

single, prominent ridge from blowholes to end of snout

HEAD

The head is one of the most distinctive features of a Minke Whale and makes it a relatively easy species to identify at close range. When observing the head from above, look for the single, sharp longitudinal ridge along the center and the overall triangular shape; from the side, look for the flat shape of the rostrum. The narrow, pointed snout of the Minke is especially distinctive, both from the side and from above. The double blowhole is typical of all baleen whales.

narrow, pointed snout

triangular head

HEAD
(FROM ABOVE)

DIVE SEQUENCE
1. The sharply pointed snout usually breaks the surface first and at a slight angle.

2. The head drops to a much shallower angle as the blowholes appear (the blow begins well before the blowholes reach the surface).

Group size 1 (1–3), occasionally up to 100 or more at good feeding areas	Fin position Far behind center

Status Common	Population *c.*500,000–1 million	Threats 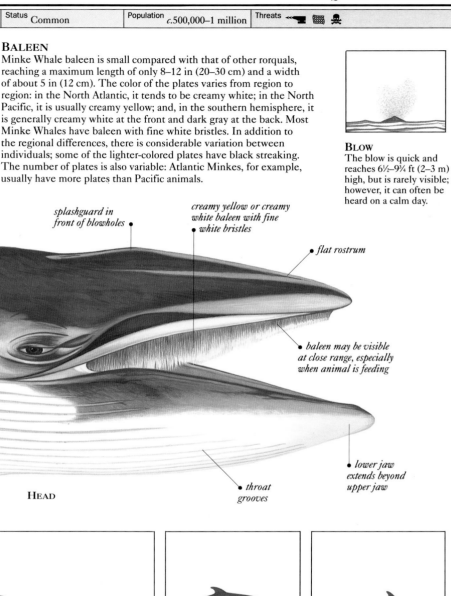

BALEEN
Minke Whale baleen is small compared with that of other rorquals, reaching a maximum length of only 8–12 in (20–30 cm) and a width of about 5 in (12 cm). The color of the plates varies from region to region: in the North Atlantic, it tends to be creamy white; in the North Pacific, it is usually creamy yellow; and, in the southern hemisphere, it is generally creamy white at the front and dark gray at the back. Most Minke Whales have baleen with fine white bristles. In addition to the regional differences, there is considerable variation between individuals; some of the lighter-colored plates have black streaking. The number of plates is also variable: Atlantic Minkes, for example, usually have more plates than Pacific animals.

BLOW
The blow is quick and reaches 6½–9¾ ft (2–3 m) high, but is rarely visible; however, it can often be heard on a calm day.

splashguard in front of blowholes

creamy yellow or creamy white baleen with fine white bristles

flat rostrum

baleen may be visible at close range, especially when animal is feeding

HEAD

throat grooves

lower jaw extends beyond upper jaw

3. The blowholes and dorsal fin are often visible simultaneously, distinguishing the Minke from all similar rorquals except the Sei Whale.

4. The back and tail stock begin to arch (much more strongly than in the Sei Whale) in preparation for a long dive.

5. The tail stock is strongly arched as the animal dives, but the flukes do not appear above the surface.

Birth wt *c.*770 lb (350 kg)	Adult wt 5–10 tons	Diet

Family BALAENOPTERIDAE	Species *Balaenoptera borealis*	Habitat 〜〜

SEI WHALE

The Sei Whale is less well known than the other members of the rorqual family. It closely resembles Bryde's Whale (p.64) in both size and appearance, and for many years the two species were frequently confused. From a distance, it is almost impossible to tell them apart, although differences in their dive sequences, head ridges, and distribution can be useful for identification. Confusion is also possible with Fin Whales (p.72) and, to a lesser extent, Minke Whales (p.56) and Blue Whales (p.68). Northern and southern hemisphere Sei Whales may belong to separate subspecies: there are subtle differences in the number of throat grooves and baleen plates. Southern animals are also slightly larger than the ones in the north, growing to a maximum length of 70 ft (21 m), compared with just over 59 ft (18 m); however, the average length for both is considerably less. Sei Whales were heavily exploited by the whaling industry, especially during the 1960s and early 1970s, and the population has been severely depleted.

• **OTHER NAMES** Pollack Whale, Coalfish Whale, Sardine Whale, Japan Finner, Rudolphi's Rorqual.

predominantly bluish gray, dark gray, or black body may appear
• *brownish in certain lights*

slightly arched
• *head*

single longitudinal
• *ridge on head*

both sides of mouth same color •

paler gray or grayish white area on throat grooves •

32–62 throat grooves, usually end just behind flippers •

slender, relatively short flippers, one-tenth of body length

• *pointed tips*

• *dark upper sides and undersides of flippers*

BALEEN
300–410 each side

BEHAVIOR
More regular dive sequence than most other rorquals and stays near the surface more consistently. Normally blows once every 40 to 60 seconds, though may blow every 20 to 30 seconds for 1 to 4 minutes, and then dive for 5 to 20 minutes. During shorter dives, rarely descends deeper than a few feet, so its progress can be followed by "fluke prints" or swirls left by the beat of the tail just below the surface. Seldom breaches. Dorsal fin and back remain visible for longer periods of time than with other large whales. Swimming behavior less erratic than Bryde's Whale, but capable of great speed.

ID CHECKLIST
• longitudinal ridge on head
• tall, sickle-shaped fin
• does not arch tail stock
• rarely shows flukes
• both sides of head evenly dark
• relatively low blow
• blowholes and fin visible simultaneously
• often swims close to surface

Group size 2–5 (1–5), larger groups of up to 30 at good feeding grounds	Fin position Far behind center

Status Locally common	Population *c.*40,000–60,000	Threats Unknown

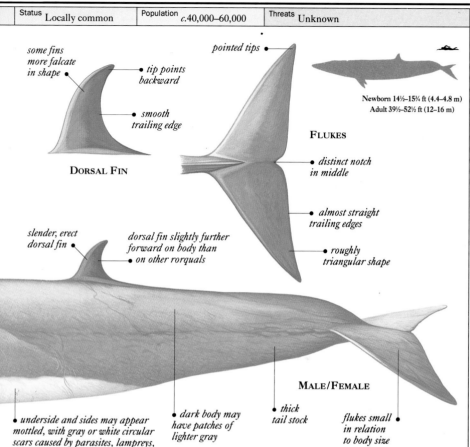

some fins more falcate in shape

tip points backward

pointed tips

Newborn 14½–15¾ ft (4.4–4.8 m)
Adult 39½–52½ ft (12–16 m)

smooth trailing edge

FLUKES

DORSAL FIN

distinct notch in middle

almost straight trailing edges

slender, erect dorsal fin

dorsal fin slightly further forward on body than on other rorquals

roughly triangular shape

MALE/FEMALE

thick tail stock

underside and sides may appear mottled, with gray or white circular scars caused by parasites, lampreys, or Cookie-cutter Sharks

dark body may have patches of lighter gray

flukes small in relation to body size

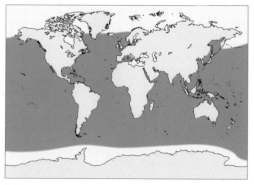

WORLDWIDE DISTRIBUTION, BUT PRIMARILY IN DEEP, TEMPERATE WATERS

WHERE TO LOOK
Less easy to predict occurrence at specific localities than most other rorquals, though there tend to be sporadic annual invasions known as "Sei Whale years," in particular areas. Not normally found in extreme polar waters, although subarctic and subantarctic are favored summer feeding grounds; believed to migrate into warmer, lower latitudes for the winter. Migrations are poorly known and probably quite irregular. There appears to be little or no mixing between northern and southern hemisphere populations. It is most common in the southern hemisphere. May be seen around islands but rarely close to shore elsewhere.

Birth wt *c.*1,600 lb (725 kg)	Adult wt 20–30 tons	Diet

Family BALAENOPTERIDAE	Species *Balaenoptera borealis*	Habitat 〰〰

BREACHING
Sei Whales do occasionally breach, but not as frequently as most other rorquals. There are relatively few documented records, though they appear to leap out of the water at a very low angle in relation to the surface and do a belly flop on the water, very quickly disappearing from view. A single breach appears to be the norm, though multiple breaches may be seen occasionally.

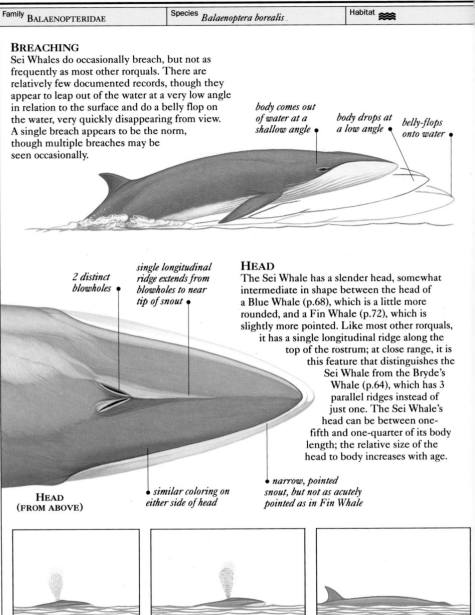

body comes out of water at a shallow angle •

body drops at a low angle •

belly-flops onto water •

2 distinct blowholes •

single longitudinal ridge extends from blowholes to near tip of snout •

HEAD
The Sei Whale has a slender head, somewhat intermediate in shape between the head of a Blue Whale (p.68), which is a little more rounded, and a Fin Whale (p.72), which is slightly more pointed. Like most other rorquals, it has a single longitudinal ridge along the top of the rostrum; at close range, it is this feature that distinguishes the Sei Whale from the Bryde's Whale (p.64), which has 3 parallel ridges instead of just one. The Sei Whale's head can be between one-fifth and one-quarter of its body length; the relative size of the head to body increases with age.

HEAD
(FROM ABOVE)

• *similar coloring on either side of head*

• *narrow, pointed snout, but not as acutely pointed as in Fin Whale*

DIVE SEQUENCE
1. The head usually rises at a shallow angle, but rises more steeply when being chased.

2. The head, major portion of the back, and sometimes the dorsal fin may break the surface together. The narrow blow is visible.

3. The blowhole and dorsal fin are visible simultaneously, distinguishing the Sei Whale from other similar rorquals, except the Minke Whale.

Group size 2–5 (1–5), larger groups of up to 30 at good feeding grounds	Fin position Far behind center

Status Locally common	Population c.40,000–60,000	Threats Unknown

BALEEN

Sei Whales in the northern hemisphere have 318 to 340 baleen plates on each side of the upper jaw, while southern hemisphere animals have 300 to 410. They reach a length of about 30–32 in (75–80 cm). The baleen is usually gray-black all over (though often with a greenish or bluish metallic sheen); in some individuals, a small number of plates near the tip of the snout are partly white or cream-colored, or streaked with white. The baleen bristles are noticeably silky in texture (possibly because the whales skim for their prey rather than lunging or gulping) and have a whitish fringe; Sei Whales have 35 to 60 bristles per half inch (1 cm), whereas all other rorquals have fewer than 35.

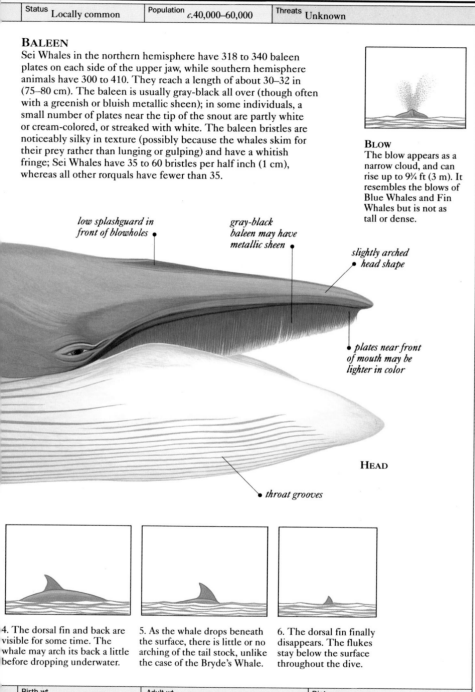

BLOW
The blow appears as a narrow cloud, and can rise up to 9¾ ft (3 m). It resembles the blows of Blue Whales and Fin Whales but is not as tall or dense.

low splashguard in front of blowholes

gray-black baleen may have metallic sheen

slightly arched head shape

plates near front of mouth may be lighter in color

HEAD

throat grooves

4. The dorsal fin and back are visible for some time. The whale may arch its back a little before dropping underwater.

5. As the whale drops beneath the surface, there is little or no arching of the tail stock, unlike the case of the Bryde's Whale.

6. The dorsal fin finally disappears. The flukes stay below the surface throughout the dive.

Birth wt c.1,600 lb (725 kg)	Adult wt 20–30 tons	Diet

Family BALAENOPTERIDAE	Species *Balaenoptera edeni*	Habitat 🐋 🌊

BRYDE'S WHALE

Bryde's Whale is strikingly similar to the Sei Whale (p.60) in both size and appearance, which has often led to confusion between the 2 species. At long range, only 1 or 2 distinguishing features may be visible: Bryde's surfaces and blows less regularly and, unlike the Sei, often arches its tail stock before a dive. Confusion is also possible with Minke Whales (p.56) and Fin Whales (p.72). However, the Bryde's Whale is unique in having 3 longitudinal ridges on its head, while all other members of the family have just one. There may be at least 2 distinct forms in some areas: one occurs offshore and is partly migratory, the other lives inshore and may be resident year-round. The 2 forms differ slightly in their reproductive behavior, and offshore animals are usually larger, have more scarring, and have longer and broader baleen than inshore animals. There may also be a "dwarf" form living around the Solomon Islands, in the Pacific Ocean. Some populations have been depleted by whaling. The population figure of 90,000 is a very rough estimate.

• **OTHER NAME** Tropical Whale.

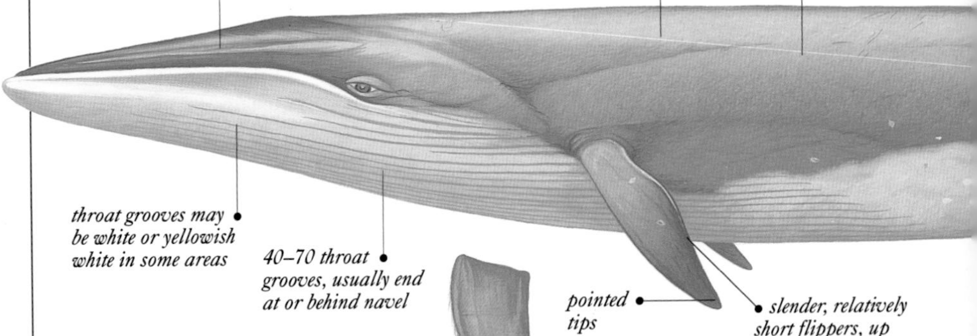

3 parallel longitudinal • ridges on head

smoky gray body color may appear chocolate brown or golden in certain lights •

• slender body

throat grooves may • be white or yellowish white in some areas

40–70 throat • grooves, usually end at or behind navel

pointed • tips

• slender, relatively short flippers, up to one-tenth of body length

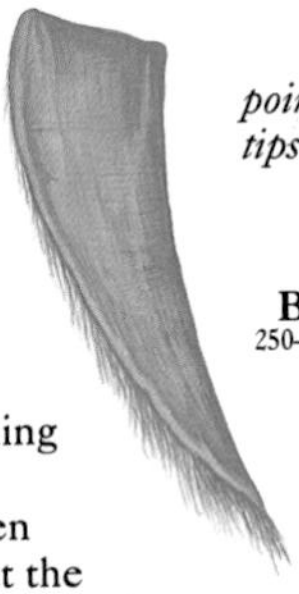

BALEEN
250–365 each side

BEHAVIOR

Poorly known. Occasionally inquisitive and will approach boats, circling them or swimming alongside. Sometimes breaches clear of the water. When feeding, typically makes sudden changes in direction, both underwater and at the surface; its swimming style often gives the impression of a large dolphin rather than a whale. Feeds year-round. Breathing sequence seldom regular but averages 4 to 7 blows followed by a long dive of up to 8 minutes (though normally less than 2 minutes); however, capable of staying underwater for longer. Loose aggregations may be spread over several square miles. When surfacing between short dives, rarely shows more than top of head; however, back and dorsal fin usually visible just before a long dive.

ID CHECKLIST

- 3 parallel ridges on head
- prominent, falcate fin
- skin may be mottled
- dark upper side
- tall, thin, hazy blow
- may arch tail stock on diving
- flukes rarely seen above surface
- irregular dive sequence
- often inquisitive

Group size 1–2 (1–7), loose groups of up to 30 at good feeding grounds	Fin position Far behind center

Status Locally common	Population *c.*90,000	Threats Unknown

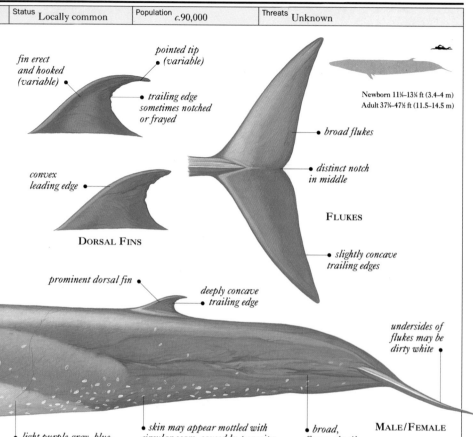

fin erect and hooked (variable)

pointed tip (variable)

trailing edge sometimes notched or frayed

Newborn 11¼–13¼ ft (3.4–4 m)
Adult 37¾–47½ ft (11.5–14.5 m)

broad flukes

distinct notch in middle

convex leading edge

FLUKES

DORSAL FINS

slightly concave trailing edges

prominent dorsal fin

deeply concave trailing edge

undersides of flukes may be dirty white

light purple-gray, blue-gray, or creamy gray underside

skin may appear mottled with circular scars, caused by parasites or Cookie-cutter Sharks

broad, flattened tail stock

MALE/FEMALE

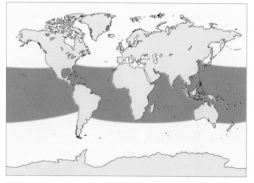

WORLDWIDE IN TROPICAL, SUBTROPICAL, AND SOME WARM
TEMPERATE WATERS

WHERE TO LOOK
Known to occur between 40° N and 40° S, and may extend into higher latitudes where there are warm water currents. However, prefers water temperatures above 68° F (20° C), so most common in tropical and subtropical areas, between 30° N and 30° S. Distribution may not be continuous throughout the range, and there appear to be local pockets of abundance, such as off South Africa, Japan, Sri Lanka, Fiji, and western Australia. Offshore animals may migrate short distances, but there are no known long-distance migrations to higher latitudes. Earlier records of distribution confused by misidentification with Sei Whale.

Birth wt 1,985 lb (900 kg)	Adult wt 12–20 tons	Diet

Family BALAENOPTERIDAE	Species *Balaenoptera edeni*	Habitat 〰️ 〰️

BREACHING

Breaching is commonly observed in Bryde's Whales in some areas but rarely in others. It often follows short periods of strenuous activity such as high-speed swimming. In exceptional cases, a single animal may breach dozens of times in a row (one individual in Japan, off the coast of Ogata, was observed doing 70 breaches nonstop); but 2 or 3 times in a row is more typical. The whale normally leaves the water almost vertically, at an angle of between 70° and 90°, sometimes arching its back in midair, and either simply falls into the water or twists first. Some animals clear the surface completely, but usually the rear quarter of the body (roughly as far as the dorsal fin) remains out of sight underwater.

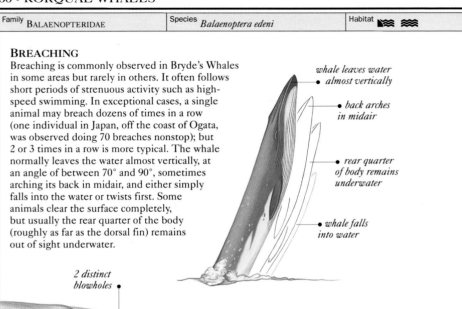

whale leaves water almost vertically

back arches in midair

rear quarter of body remains underwater

whale falls into water

2 distinct blowholes

HEAD

At close range, the Bryde's Whale is unmistakable because of the 3 longitudinal ridges on the top of its head. All other members of the rorqual family have just one. All 3 ridges are normally ½–¾ in (1–2 cm) high, but the prominence of the 2 additional outer ones is variable and, in some individuals, one or both of them may be difficult to detect at sea. The outer ridges do not quite reach the tip of the snout and, near the blowholes, they disappear from the surface and change into grooves of varying lengths. The grooves are sometimes absent in some individuals. The central ridge is continuous.

HEAD
(FROM ABOVE)

central ridge flanked by 2 additional parallel ridges

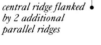

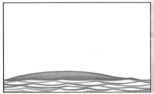

DIVE SEQUENCE
1. The snout breaks the surface at a low angle (more steeply after a deep dive).

2. The whale blows (this may not be seen from a distance) and straightens out its body. The mouth line may be visible.

3. The blowholes usually disappear from sight just before the dorsal fin becomes visible. The long back remains low in the water.

Group size 1–2 (1–7), loose groups of up to 30 at good feeding grounds	Fin position Far behind center

Status Locally common	Population *c*.90,000	Threats Unknown

BALEEN

The baleen of the Bryde's Whale is fairly distinctive in shape. It is short and wide, reaching a maximum length (excluding bristles) of about 20 in (50 cm) and a width of about 7½ in (19 cm), and has a slightly concave inner margin. The number of well-developed plates normally ranges from 250 to 280, but there are also many rudimentary plates bringing the total up to a maximum of 365. There may be a gap between the plates at the front of the rostrum. Their color varies greatly between individuals; most are black or slate gray, though the plates near the tip of the rostrum are often completely or partially creamy white (sometimes with gray stripes). The bristles are long, stiff, and uncurled, and are generally brownish or grayish in color.

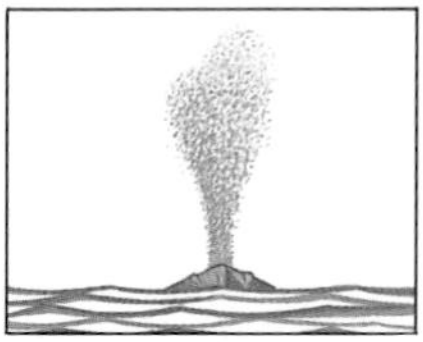

BLOW
The blow is tall and thin, rising 9¾–13¼ ft (3–4 m) high in a single cloud; often it is not clearly visible from a distance.

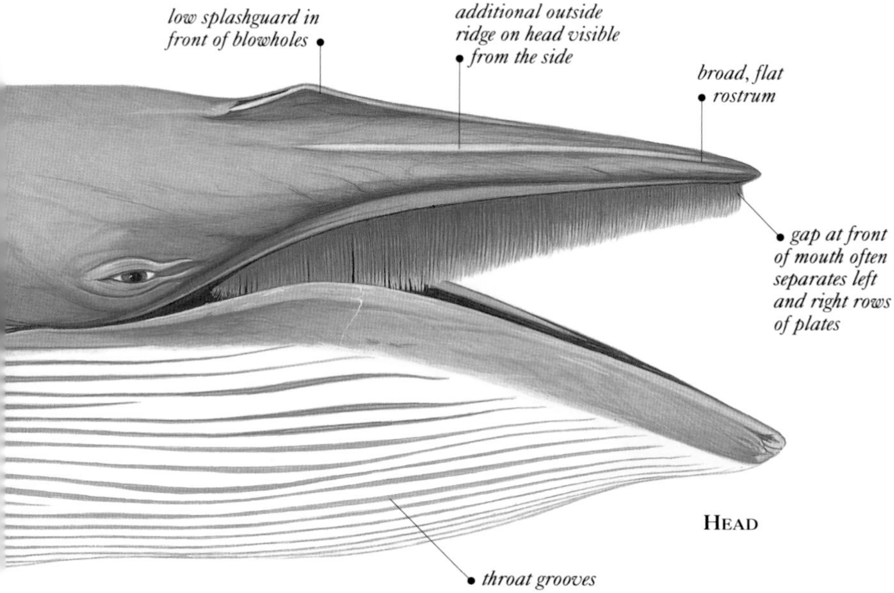

low splashguard in front of blowholes •

additional outside ridge on head visible • *from the side*

broad, flat • *rostrum*

• *gap at front of mouth often separates left and right rows of plates*

HEAD

• *throat grooves*

4. The dorsal fin usually appears just after the blowholes have dropped below the surface.

5. The whale begins to roll sharply forward and the back arches in preparation for a deep dive.

6. The tail stock is arched strongly before a dive, unlike in the case of the Sei Whale.

7. The flukes rarely (if ever) appear above the surface when the whale dives.

Birth wt 1,985 lb (900 kg)	Adult wt 12–20 tons	Diet

Family BALAENOPTERIDAE	Species *Balaenoptera musculus*	Habitat 〰〰 (🐋〰)

BLUE WHALE

The Blue Whale is one of the largest animals ever to have lived on earth. A length of more than 110 ft (33 m) and a weight of around 190 tons have been recorded, but the average size is much smaller. There are believed to be 3 different subspecies: *Balaenoptera musculus* subsp. *intermedia* in the southern hemisphere, the slightly smaller subsp. *musculus* in the northern hemisphere, and the even smaller subsp. *brevicauda* (the Pygmy Blue Whale), which occurs mainly in tropical areas of the southern hemisphere. Despite a number of fairly subtle differences, the Pygmy Blue may be impossible to distinguish from the two larger Blue Whales at sea. All three subspecies can be confused with Sei Whales (p.60) or Fin Whales (p.72), especially at a distance. The Blue Whale was hunted close to extinction by the whaling industry: the mortality rates were so high that some populations may never recover.

• **OTHER NAMES** Sulfur-bottom, Sibbald's Rorqual, Great Northern Rorqual.

broad, flattened head •

large splashguard •

raised ridge along spine behind head (variable) •

variable pale gray or white mottling, mainly behind head •

pale blue-gray body color (variable) •

• *both sides of mouth uniformly blue-gray*

55–88 throat grooves, usually end at or behind navel •

long, slender flippers, up to one-seventh of body length

BALEEN
270–395 each side

• *pale blue-gray or white undersides*

• *pointed tips may be lighter than rest of flippers*

BEHAVIOR

Blowing and diving patterns vary according to whale's activity. When relaxed, blows every 10 to 20 seconds for a total of 2 to 6 minutes, and then dives for 5 to 20 minutes (able to stay under longer). Probably dives to depths of up to 490 ft (150 m), but can go deeper. Can accelerate to speeds of over 19 mph (30 km/h) when being chased, but usually much slower. Some individuals are easy to approach, while others can be difficult. Adults rarely, if ever, breach clear of the water; however, youngsters have been observed breaching, usually at an angle of about 45°, and landing on their stomachs or sides. In some areas, most feeding seems to take place during the evening and early morning.

ID CHECKLIST

• enormous size
• blue-gray body color
• mottled appearance
• tiny, stubby fin set far back
• broad, flattened, U-shaped head
• huge blowhole splashguard
• extremely thick tail stock
• 29½-ft (9-m) high blow
• may show flukes on diving

Group size 1–2 (1–5), sometimes larger gatherings at feeding grounds	Fin position Far behind center

Status Endangered	Population *c.*6,000–14,000	Threats Unknown

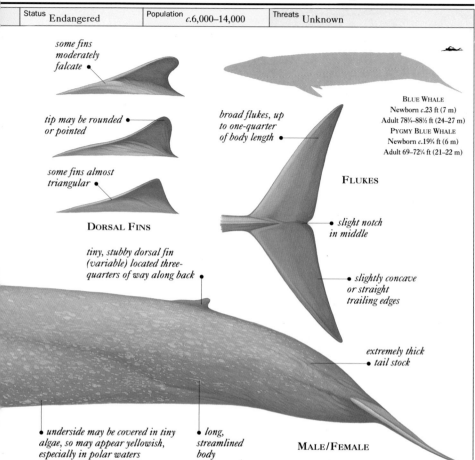

*some fins
moderately
falcate* •

*tip may be rounded
or pointed* •

*broad flukes, up
to one-quarter
of body length* •

BLUE WHALE
Newborn *c.*23 ft (7 m)
Adult 78¾–88½ ft (24–27 m)
PYGMY BLUE WHALE
Newborn *c.*19¾ ft (6 m)
Adult 69–72¼ ft (21–22 m)

*some fins almost
triangular* •

FLUKES

DORSAL FINS

• *slight notch
in middle*

*tiny, stubby dorsal fin
(variable) located three-
quarters of way along back* •

• *slightly concave
or straight
trailing edges*

*extremely thick
• tail stock*

• *underside may be covered in tiny
algae, so may appear yellowish,
especially in polar waters*

• *long,
streamlined
body*

MALE/FEMALE

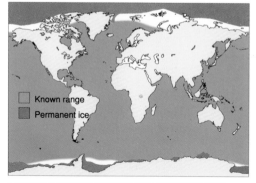

☐ Known range
▨ Permanent ice

PATCHILY DISTRIBUTED WORLDWIDE, MAINLY IN COLD
WATERS AND OPEN SEAS

WHERE TO LOOK

Three main populations are recognized:
in the North Atlantic, North Pacific, and
southern hemisphere. Distribution not
continuous across range. Most live in
the southern hemisphere, but often
seen in parts of California; Gulf of
California (Sea of Cortez), Mexico; Gulf
of St. Lawrence, Canada; and northern
Indian Ocean. Only a few hundred left
in the North Atlantic. May migrate long
distances between low-latitude wintering
grounds and high-latitude summering
grounds. Population in the northern Indian
Ocean may be resident year-round. Mainly
found along edge of the continental shelf
and near polar ice.

Birth wt *c.*2.5 tons	Adult wt 100–120 tons	Diet 🦐

Family BALAENOPTERIDAE	Species *Balaenoptera musculus*	Habitat 〰〰 (🐋〰)

UNDERSIDE

The underside of a Blue Whale may appear to be yellowish or mustard-colored (it is for this reason that the Blue Whale is sometimes known as the "Sulfur-bottom"). The yellowish color is not true pigmentation but is caused by the presence of microscopic algae, called diatoms, which attach themselves to the whale's body. This is most commonly observed in animals living in cold waters near the poles.

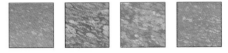

COLOR VARIATIONS (UPPER SIDE AND SIDES)

UPPER SIDE AND SIDES

Blue Whales show considerable variation in color between individuals. They are all basically blue-gray, but the true color varies from a uniform dark slate gray, with relatively little white mottling, to a very light blue, with extensive mottling. Pygmy Blue Whales may be lighter in color than the larger Blue Whales although, again, there seems to be considerable variation.

COLOR VARIATIONS (UNDERSIDE)

HEAD

The head of the Blue Whale is very distinctive. It is long, forming up to one-quarter of the total body length, and very broad compared with those of other rorquals. Viewed from above, it is basically U-shaped, though it is often likened to the shape of a Gothic arch; as in most other rorquals, it has a single longitudinal ridge along the top of the rostrum. The exceptionally large, fleshy splashguard, which surrounds the blowholes at the front and sides, is the most prominent feature.

2 distinct blowholes • *flat rostrum* • *broad, U-shaped head* •

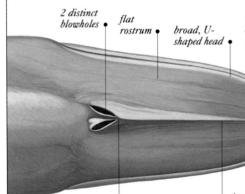

HEAD
(FROM ABOVE)

• *raised splashguard*

• *single longitudinal ridge runs from blowholes to near tip of snout*

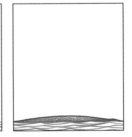

DIVE SEQUENCE
1. When swimming slowly, the whale rises at a shallow angle.

2. The whale blows as soon as the head begins to break the surface. The blow is tall and upright.

3. The head disappears below the surface, and a long expanse of the back rolls into view.

4. The dorsal fin normally appears some time after the blow has dispersed and the head has disappeared.

Group size 1–2 (1–5), sometimes larger gatherings at feeding grounds	Fin position Far behind center

Status Endangered	Population *c.*6,000–14,000	Threats Unknown

BALEEN

The Blue Whale has the longest baleen of all the rorqual whales. However, the plates are wide in relation to their length – they can be 20–22 in (50–55 cm) wide and 35–39 in (90 cm–1 m) long – and are roughly triangular in shape; the baleen is smaller in the Pygmy Blue Whale. The stiff plates and palate are usually uniformly jet black or blue-black in color, although there may be considerable individual variation; the coarse bristles of the baleen are sometimes gray in older animals.

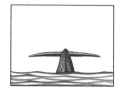

FLUKING

The flukes are lifted out of the water briefly on some dives only; they are usually raised at an angle shallower than 45°.

BLOW

The blow is spectacular and appears as a slender, vertical column of spray rising to 29½ ft (9 m) high. It can range from 19¾ to 39½ ft (6–12 m) high.

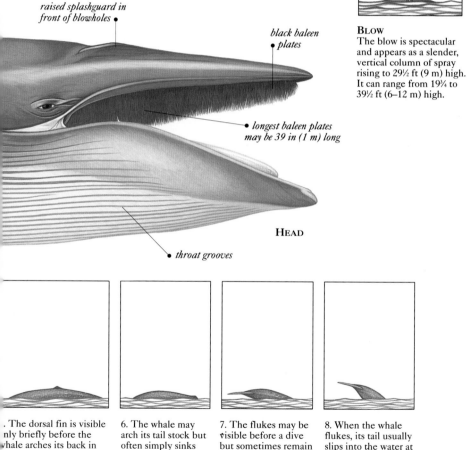

raised splashguard in front of blowholes

black baleen plates

longest baleen plates may be 39 in (1 m) long

HEAD

throat grooves

. The dorsal fin is visible nly briefly before the whale arches its back in reparation for the dive.

6. The whale may arch its tail stock but often simply sinks below the surface.

7. The flukes may be visible before a dive but sometimes remain below the surface.

8. When the whale flukes, its tail usually slips into the water at a shallow angle.

Birth wt *c.*2.5 tons	Adult wt 100–120 tons	Diet

Family BALAENOPTERIDAE	Species *Balaenoptera physalus*	Habitat 〜〜〜 ▰〜

FIN WHALE

The Fin Whale is the second largest animal on earth (after the Blue Whale). It is known to grow to more than 85¼ ft (26 m), though the average length is much smaller. Animals in the northern hemisphere are typically 39 in–5 ft (1–1.5 m) smaller than those in the southern hemisphere; some authorities recognize them as separate subspecies. It is most likely to be confused with the Sei Whale (p.60) or Blue Whale (p.68) or, in the tropics, Bryde's Whale (p.64). An invaluable aid to identifying the Fin Whale at close range is the asymmetrical

pigmentation on its head: on the right side, the lower "lip," mouth cavity, and some of the baleen plates are white; the left side is uniformly gray. When swimming just below the surface, the white "lip" is often clearly visible, though it could be confused with the white flipper of a Humpback Whale (p.76). Once one of the most abundant of the large whales, the Fin Whale was heavily exploited by the whaling industry and its population has been severely depleted.
• **OTHER NAMES** Finback, Finner, Common Rorqual, Razorback, Herring Whale.

tip of rostrum level
• (no downward turn)

variable grayish white chevron
behind head on each side
(more prominent on right) •

silvery gray, dark
gray, or brownish
black body color •

long,
streamlined
body •

• lower "lip"
dark on left side,
but white on right

56–100 throat
grooves, usually end
at or behind navel •

white •
underside

• slender,
relatively
short flippers

BALEEN
260–480 each side

FLIPPER
UNDERSIDE

pointed
• tip

BEHAVIOR
Neither avoids boats nor approaches them. It is almost impossible to judge when it will surface or how far away: obtaining a close-up view can be difficult. Surfacing motion depends on whether the whale is moving leisurely at the surface or surfacing from a deep dive. It typically blows 2 to 5 times, at intervals of 10 to 20 seconds, before diving for 5 to 15 minutes (though can stay under for longer). Dives to depths of at least 755 ft (230 m). Asymmetrical pigmentation may be linked with the way the whale swims on its right side while feeding. It sometimes breaches clear of the water. Fast swimmer, capable of reaching speeds over 19 mph (30 km/h). More commonly seen in small groups than other rorqual whales.

ID CHECKLIST
- exceptionally large size
- asymmetrical head pigmentation
- small, backward-sloping fin
- longitudinal ridge on head
- tall, narrow blow
- fin visible soon after blow
- rarely shows flukes
- grayish white chevron
- indifferent to boats

Group size 3–7 (1–2), 100 or more may gather at good feeding grounds	Fin position Far behind center

Status	Population	Threats
Locally common	*c.*120,000	☠

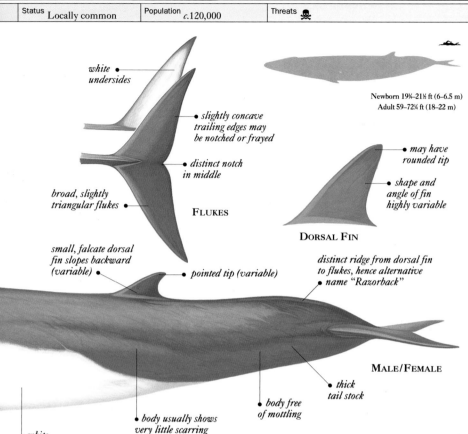

white undersides

slightly concave trailing edges may be notched or frayed

distinct notch in middle

broad, slightly triangular flukes

FLUKES

Newborn 19¾–21½ ft (6–6.5 m)
Adult 59–72¼ ft (18–22 m)

may have rounded tip

shape and angle of fin highly variable

DORSAL FIN

small, falcate dorsal fin slopes backward (variable)

pointed tip (variable)

distinct ridge from dorsal fin to flukes, hence alternative name "Razorback"

MALE/FEMALE

thick tail stock

body free of mottling

body usually shows very little scarring

white underside

WHERE TO LOOK

Most common in the southern hemisphere, least common in the tropics. Will enter polar waters, but not as often as Blue Whales or Minke Whales. Only rorqual commonly found in the Mediterranean. There are probably 3 geographically isolated populations: in the North Atlantic, North Pacific, and southern hemisphere. Some populations may migrate between relatively warm low latitudes in winter and cooler high latitudes in summer, though movements less predictable than in some other large whales. Certain lower latitude populations, such as in the Gulf of California (Sea of Cortez), Mexico, seem to be resident year-round. Usually occurs in offshore waters but may be seen close to shore in areas where water is deep enough.

☐ Known range
■ Permanent ice

WORLDWIDE DISTRIBUTION, BUT MOST COMMON IN
TEMPERATE WATERS AND IN THE SOUTHERN HEMISPHERE

Birth wt	Adult wt	Diet
*c.*2 tons	30–80 tons	

Family BALAENOPTERIDAE	Species *Balaenoptera physalus*	Habitat 〜〜〜 ▨〜

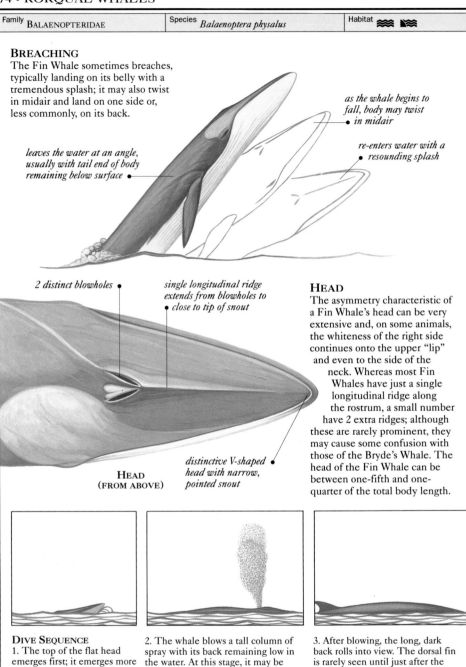

BREACHING
The Fin Whale sometimes breaches, typically landing on its belly with a tremendous splash; it may also twist in midair and land on one side or, less commonly, on its back.

as the whale begins to fall, body may twist • in midair

leaves the water at an angle, usually with tail end of body remaining below surface •

re-enters water with a • resounding splash

2 distinct blowholes •

single longitudinal ridge extends from blowholes to • close to tip of snout

distinctive V-shaped • head with narrow, pointed snout

HEAD
(FROM ABOVE)

HEAD
The asymmetry characteristic of a Fin Whale's head can be very extensive and, on some animals, the whiteness of the right side continues onto the upper "lip" and even to the side of the neck. Whereas most Fin Whales have just a single longitudinal ridge along the rostrum, a small number have 2 extra ridges; although these are rarely prominent, they may cause some confusion with those of the Bryde's Whale. The head of the Fin Whale can be between one-fifth and one-quarter of the total body length.

DIVE SEQUENCE
1. The top of the flat head emerges first; it emerges more steeply after a deeper dive.

2. The whale blows a tall column of spray with its back remaining low in the water. At this stage, it may be possible to see the white right side.

3. After blowing, the long, dark back rolls into view. The dorsal fin is rarely seen until just after the final blow before a long dive.

Group size 3–7 (1–2), 100 or more may gather at good feeding grounds	Fin position Far behind center

Status	Population	Threats
Locally common	c.120,000	☠

BALEEN PLATES

Fin Whale baleen reaches a maximum length of about 28–35 in (70–90 cm) and a width of 8–12 in (20–30 cm). It is unusual in having asymmetrical pigmentation, as on the head: on the right side, the plates in the front quarter to third of the mouth are white, creamy white, or yellowish white in color; the remaining plates on the right side, and all of the ones on the left side, are dark gray (often with alternating yellowish white and bluish gray vertical stripes). The baleen bristles are soft compared with those of the Blue Whale and vary from yellowish white to grayish white.

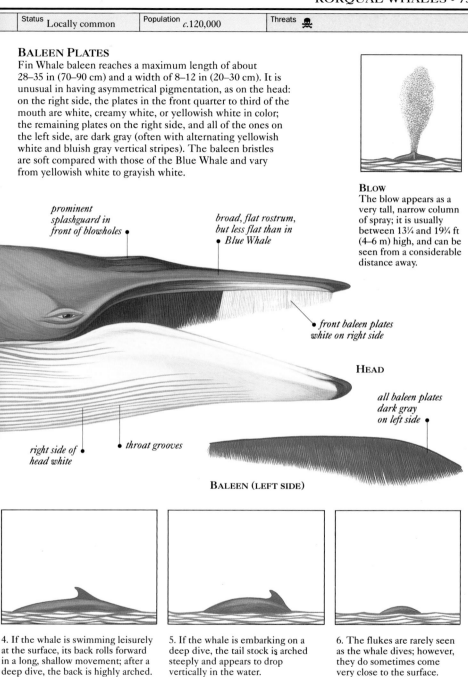

BLOW
The blow appears as a very tall, narrow column of spray; it is usually between 13¼ and 19¾ ft (4–6 m) high, and can be seen from a considerable distance away.

prominent splashguard in front of blowholes •

broad, flat rostrum, but less flat than in • Blue Whale

• front baleen plates white on right side

HEAD

all baleen plates dark gray on left side •

right side of • head white

• throat grooves

BALEEN (LEFT SIDE)

4. If the whale is swimming leisurely at the surface, its back rolls forward in a long, shallow movement; after a deep dive, the back is highly arched.

5. If the whale is embarking on a deep dive, the tail stock is arched steeply and appears to drop vertically in the water.

6. The flukes are rarely seen as the whale dives; however, they do sometimes come very close to the surface.

Birth wt	Adult wt	Diet
c.2 tons	30–80 tons	🦐 🐟 (◄●)

Family BALAENOPTERIDAE	Species *Megaptera novaeangliae*	Habitat 〰️ 〰️

HUMPBACK WHALE

One of the most energetic of the large whales, the Humpback Whale is well known for its spectacular breaching, lobtailing, and flipper-slapping. It is also one of the easiest whales to identify. At a distance, it can usually be distinguished by its unique flukes; at close range, its knobbly head and long flippers are unmistakable. However, no two Humpbacks are exactly alike: the black and white pigmentation on the undersides of their flukes is as unique as a human fingerprint; as a result, experts have been able to distinguish and name thousands of individuals around the world. Males at their breeding grounds are well known for singing the longest and most complex songs in the animal kingdom. More than 100,000 Humpbacks were killed by whalers and, although some stocks seem to be recovering, today's population is just a fraction of its original size.

• OTHER NAME Hump-backed Whale.

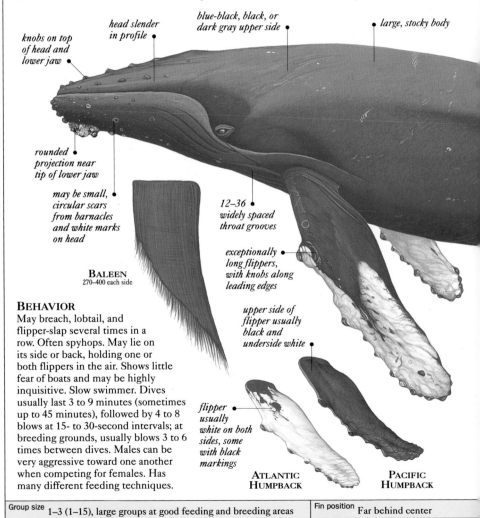

knobs on top of head and lower jaw •

head slender in profile •

blue-black, black, or dark gray upper side •

• large, stocky body

• rounded projection near tip of lower jaw

may be small, circular scars from barnacles and white marks on head

12–36 • widely spaced throat grooves

exceptionally • long flippers, with knobs along leading edges

BALEEN
270–400 each side

upper side of flipper usually black and underside white •

BEHAVIOR
May breach, lobtail, and flipper-slap several times in a row. Often spyhops. May lie on its side or back, holding one or both flippers in the air. Shows little fear of boats and may be highly inquisitive. Slow swimmer. Dives usually last 3 to 9 minutes (sometimes up to 45 minutes), followed by 4 to 8 blows at 15- to 30-second intervals; at breeding grounds, usually blows 3 to 6 times between dives. Males can be very aggressive toward one another when competing for females. Has many different feeding techniques.

flipper • usually white on both sides, some with black markings

ATLANTIC HUMPBACK

PACIFIC HUMPBACK

Group size 1–3 (1–15), large groups at good feeding and breeding areas	Fin position Far behind center

Status Rare	Population 12,000–15,000	Threats

ID CHECKLIST

- black or dark gray upper side
- low, stubby fin with hump
- large, stocky body
- long, white or black flippers
- knobs on head and lower jaw
- flukes raised before deep dive
- irregular, wavy edges on flukes
- single bushy blow
- may be inquisitive

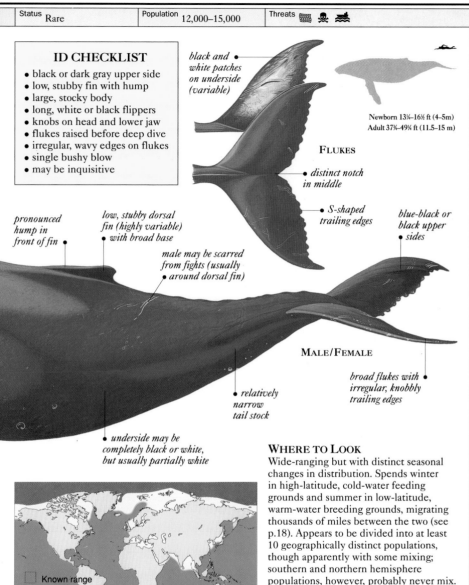

black and white patches on underside (variable)

Newborn 13¼–16½ ft (4–5m)
Adult 37¾–49¼ ft (11.5–15 m)

FLUKES

• distinct notch in middle

• S-shaped trailing edges

blue-black or black upper • sides

pronounced hump in front of fin •

low, stubby dorsal fin (highly variable) • with broad base

male may be scarred from fights (usually • around dorsal fin)

MALE/FEMALE

broad flukes with • irregular, knobbly trailing edges

• relatively narrow tail stock

• underside may be completely black or white, but usually partially white

Known range
Permanent ice

WIDELY DISTRIBUTED IN ALL OCEANS FROM THE POLES TO
THE TROPICS

WHERE TO LOOK

Wide-ranging but with distinct seasonal changes in distribution. Spends winter in high-latitude, cold-water feeding grounds and summer in low-latitude, warm-water breeding grounds, migrating thousands of miles between the two (see p.18). Appears to be divided into at least 10 geographically distinct populations, though apparently with some mixing; southern and northern hemisphere populations, however, probably never mix. Population in northern Indian Ocean may be resident year-round, or may migrate to and from Antarctica. Northeast Atlantic population as low as a few hundred. Spends much of year fairly close to continental shores or islands, breeding and feeding on shallow banks, but migrates across open seas.

Birth wt 1–2 tons	Adult wt 25–30 tons	Diet

Family BALAENOPTERIDAE	Species *Megaptera novaeangliae*	Habitat 🐋🐋

GROUP OF FIVE BLOWING
The Humpback's blow is very distinctive, although it varies in shape according to the individual, wind conditions, and the length of the previous dive.

BREACHING
Breaches vary from a full leap clear of the water to a leisurely surge with less than half of the body emerging. The whale usually lands on its back, but it sometimes emerges dorsal side up and does a belly flop; this is normally accompanied by a violent exhalation. There is some indication that breaching is more common at breeding grounds and in strong winds and that, in some areas, it peaks around midday.

whale emerges sideways •

long flippers outstretched like enormous wings

arches back and does half turn

lands on back with explosion of spray

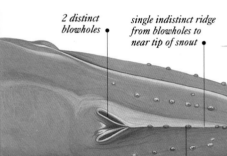

2 distinct blowholes •

single indistinct ridge from blowholes to near tip of snout •

knobs along central ridge and in haphazard arrangement elsewhere

HEAD (FROM ABOVE)

HEAD
Viewed from above, the Humpback has a broad, relatively rounded head, constituting up to one-third of its total body length. Its most distinctive feature is a series of knobs, or tubercles, which cover the rostrum (in front of the blowholes) and much of the lower jaw. These vary in number and location from one individual to another. About the size of a golf ball, each knob is a hair follicle with a single, coarse hair between ½ and 1¼ in (1–3 cm) long growing out of its center; this suggests that they may have some kind of sensory function.

DIVE SEQUENCE
1. The splashguard and blowholes appear above the surface first.

2. As the dorsal fin comes into view, the distinctive sloping back forms a shallow triangle with the surface of the sea.

3. The body arches, forming a much higher triangle and making the hump on the back especially evident.

Group size 1–3 (1–15), large groups at good feeding and breeding areas	Fin position Far behind center

Status	Population	Threats
Rare	12,000–15,000	

FEEDING

Humpbacks have developed some of the most diverse and spectacular feeding techniques of all baleen whales. They lunge through patches of krill or fish, gulping vast mouthfuls, and even stun their prey with slaps of their flippers or flukes. But their most impressive technique is "bubble-netting." They begin by swimming in a spiral beneath a shoal of fish or krill, blowing out air from their blowholes; this forms a "net" of bubbles, up to 150 ft (45 m) across, which surrounds the prey. With their mouths open, the whales then swim up through the center toward the surface. A bubble net usually appears as a circle or an arc of bubbles on the surface.

BLOW
The blow is very visible and distinctive: it is bushy, 8¼–9¾ ft (2.5–3 m) tall, and usually wide relative to its height.

FLUKING
Humpback flukes, with their knobbly trailing edges and black and white markings on the undersides, are unique.

prominent splashguard •

baleen relatively short and wide: maximum length 28–39 in (70–100 cm); • maximum width 12 in (30 cm)

• baleen usually black to olive brown, but occasionally whitish in color, and often with grayish white bristles

• rounded projection near tip of lower jaw appears to increase in size with age

• throat grooves

HEAD

4. As the dorsal fin drops below the surface, the tail stock is strongly arched, and the whale rolls into a dive.

5. The tail stock drops lower and continues to roll forward as the whale steepens the angle of its descent.

6. The flukes begin to appear above the surface as the whale dives more deeply.

7. The flukes are lifted high on most dives but may not be raised in shallow water.

Birth wt	Adult wt	Diet
1–2 tons	25–30 tons	

SPERM WHALES

A LL SPERM WHALES have a wax-filled structure known as the spermaceti organ inside their heads; its function is still a matter of debate, but it may be used to control the whale's buoyancy in the water and possibly as an acoustic lens to direct sound beams for echolocation. Despite this common characteristic, the three species are very different from one another; as a result, the Pygmy Sperm Whale and Dwarf Sperm Whale have recently been placed in a family of their own, known as the Kogiidae; they were originally in the family Physeteridae with their much larger and better-known relative, the Sperm Whale. They all prefer deep waters, where they feed primarily on squid, and are seldom encountered close to shore except in particular areas where the depth increases rapidly.

squarish head •

single blowhole •

• narrow, underslung lower jaw

▽ DWARF SPERM WHALE

CHARACTERISTICS

There are many variations in size within the group: the Dwarf Sperm Whale can be as small as 7 ft (2.1 m) and weigh as little as 300 lb (135 kg), compared to the 59-ft (18-m) male Sperm Whale – by far the largest of all toothed whales – which can weigh up to 50 tons. The size of their heads, in proportion to their bodies, also differs: it may be up to 15 percent of body length in the smaller two species, and up to 35 percent in male Sperm Whales (these have the largest heads in the animal kingdom).

lifts head without rolling body forward

drops below surface and dives

rises to surface slowly and deliberately

DIVE SEQUENCE

The dive sequence of Dwarf and Pygmy Sperm Whales is poorly known, although their surfacing behavior is characteristically slow. Unlike most other whales and dolphins, they rarely roll forward, but bob to the surface to take a breath and then simply drop out of sight. Reports suggest that they can swim in fast bursts if they are alarmed.

deep dives may reach 655–985 ft (200–300 m)

DISTINCTIVE BLOWHOLE
This aerial photograph shows the position of the Sperm Whale's single, slitlike blowhole, located on the left side of the head near the end of its snout.

SPECIES IDENTIFICATION

DWARF SPERM WHALE (p.84) *Smallest whale, with a prominent dorsal fin, squarish head, and false gill.*

PYGMY SPERM WHALE (p.82) *Difficult to distinguish from the Dwarf Sperm Whale at sea, but is slightly larger.*

• *robust body*

SPERM WHALE (p.86) *The Sperm Whale shows some features common to all members of the group, such as the spermaceti organ, an absence of functional teeth in the upper jaw, and the key features shown here. However, there are more differences than similarities, including the size of head in relation to the body, the dorsal fin shape, and the distance from the blowhole to the snout.*

SKIN
The wrinkly, prunelike skin of the Sperm Whale is unmistakable and unique to this species. The wrinkles are horizontal, and most are on the rear two-thirds of the body; they are less evident on fat animals at feeding grounds.

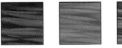

COLOR VARIATIONS
Sperm whales vary greatly in color, from dark gray to light brown; white ones, like the fictional Moby Dick, are rare.

LOGGING
Dwarf Sperm Whales and Pygmy Sperm Whales may be seen resting motionless at the surface, with part of the head and sometimes the back and dorsal fin exposed, and the tail hanging down in the water. In this state, they may be easier to approach.

Family KOGIIDAE	Species *Kogia breviceps*	Habitat 〰〰

PYGMY SPERM WHALE

The Pygmy Sperm Whale is rarely seen: it tends to live a long distance from shore and has inconspicuous habits. It is often confused with the Dwarf Sperm Whale (p.84), which was not recognized as a separate species until 1966. With so few field records, it is uncertain whether the two can be distinguished reliably except at very close range. The Pygmy Sperm Whale is most likely to be seen when resting. It floats motionless at the surface, with part of

the head and back exposed and the tail hanging down loosely in the water; animals in this state sometimes allow boats to approach closely. It may resemble a shark when it is stranded, as its underslung lower jaw and creamy white false gill are particularly apparent.
• **OTHER NAMES** Lesser Sperm Whale, Short-headed Sperm Whale, Lesser Cachalot.

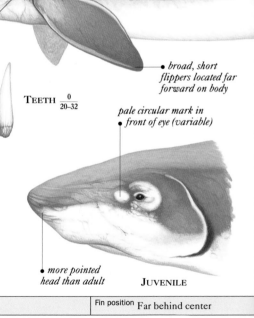

squarish head looks conical
• *when viewed from above*

distance from snout to blowhole more than one-tenth of body length •

blowhole displaced
• *slightly to left*

tiny, underslung lower jaw •

false gill •

• *broad, short flippers located far forward on body*

BEHAVIOR

Rises to the surface slowly and deliberately and, unlike most other small whales, simply drops out of sight. May occasionally breach, leaping vertically out of the water and falling back tail first or with a belly flop. Rarely approaches boats. Blow inconspicuous and low. When startled, may evacuate a reddish brown intestinal fluid and then dive, leaving behind a dense cloud in the water; this may function as a decoy, like the ink of a squid. Some records suggest that, when resting at the surface, it floats higher in the water than the Dwarf Sperm Whale.

TEETH $\frac{0}{20-32}$

pale circular mark in
• *front of eye (variable)*

• *more pointed head than adult*

JUVENILE

Group size 3–6 (1–10)	Fin position Far behind center

Status Unknown	Population Unknown	Threats Unknown

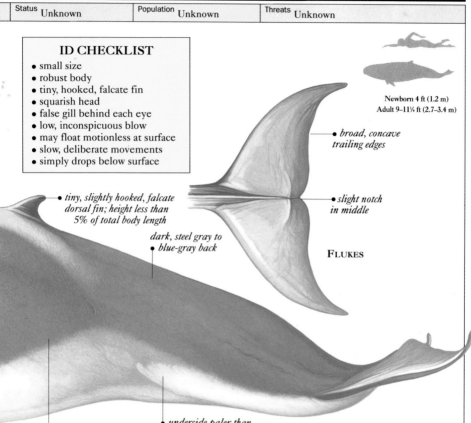

ID CHECKLIST

- small size
- robust body
- tiny, hooked, falcate fin
- squarish head
- false gill behind each eye
- low, inconspicuous blow
- may float motionless at surface
- slow, deliberate movements
- simply drops below surface

Newborn 4 ft (1.2 m)
Adult 9–11¼ ft (2.7–3.4 m)

• *broad, concave trailing edges*

• *tiny, slightly hooked, falcate dorsal fin; height less than 5% of total body length*

• *slight notch in middle*

dark, steel gray to • *blue-gray back*

FLUKES

• *body may appear wrinkled*

• *underside paler than rest of body and may be pinkish in color*

MALE / FEMALE

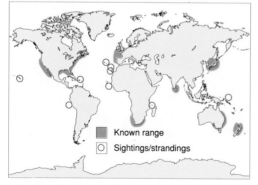

Known range

Sightings/strandings

DEEP TEMPERATE, SUBTROPICAL, AND TROPICAL WATERS
BEYOND THE CONTINENTAL SHELF

WHERE TO LOOK

Poorly known, though lack of records of live animals may be due to inconspicuous behavior rather than rarity. Most information is from strandings (especially females with calves), which may give an inaccurate picture of distribution. Seems to prefer warmer waters: there are records from nearly all temperate, subtropical, and tropical seas. Mainly a deep-water species and, unlike the Dwarf Sperm Whale, is usually seen beyond the edge of the continental shelf. Appears to be relatively common off the southeastern coast of the USA and around southern Africa, southeastern Australia, and New Zealand. Not known whether these populations are isolated.

Birth wt 120 lb (55 kg)	Adult wt 695–880 lb (315–400 kg)	Diet ◄═ (═╪═ ═╗)

Family KOGIIDAE	Species *Kogia simus*	Habitat 〰〰 (▶〰)

DWARF SPERM WHALE

The Dwarf Sperm Whale is an inconspicuous animal that generally lives a long way from shore. Rarely seen at sea except in extremely calm conditions, it is the smallest of the whales and is even smaller than some dolphins. Its square head and slow, deliberate movements distinguish it from the superficially similar Bottlenose Dolphin (p.192). However, it may be confused with the Pygmy Sperm Whale (p.82). It can be difficult or even impossible to tell the two species apart at sea, though the larger size and shape of the Dwarf Sperm Whale's dorsal fin can be distinctive. When stranded, the Dwarf Sperm Whale has a rather sharklike appearance: it has an underslung lower jaw and a creamy white arc, known as a false gill, behind each eye. The teeth in the upper jaw are vestigial.

• **OTHER NAME** Owen's Pygmy Sperm Whale.

snout to blowhole length less than one-tenth of total body length

blowhole displaced slightly to left

bluish gray or dark gray-black back

slightly pointed snout overlaps lower jaw

tiny, underslung lower jaw

false gill

flippers located far forward on body

broad, short flippers

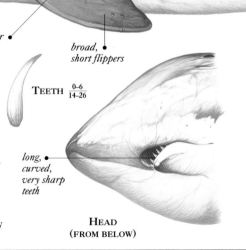

BEHAVIOR

Rises to the surface slowly and deliberately and, unlike most other small whales (which roll forward at the surface), simply drops out of sight. When startled, may evacuate a reddish brown intestinal fluid and then dive, leaving behind a dense cloud in the water; this may function as a decoy, like the ink of a squid. Probably does not approach boats. May occasionally breach, leaping vertically out of the water and falling back tail first or with a belly flop. Some records suggest that, when resting at the surface, it floats lower in the water than the Pygmy Sperm Whale. Probably dives to depths of at least 985 ft (300 m).

TEETH $\frac{0-6}{14-26}$

long, curved, very sharp teeth

HEAD
(FROM BELOW)

Group size 1–2 (1–10)	Fin position Slightly behind center

Status Unknown	Population Unknown	Threats Unknown

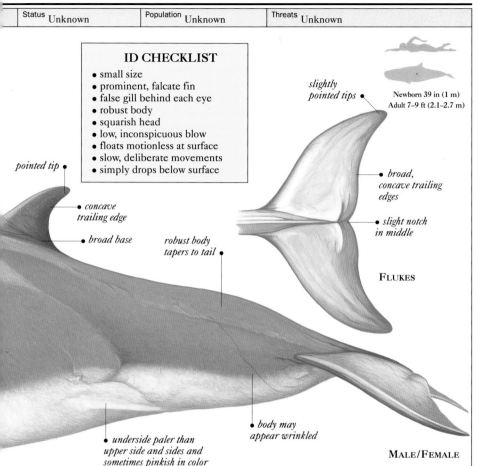

ID CHECKLIST

- small size
- prominent, falcate fin
- false gill behind each eye
- robust body
- squarish head
- low, inconspicuous blow
- floats motionless at surface
- slow, deliberate movements
- simply drops below surface

slightly pointed tips

Newborn 39 in (1 m)
Adult 7–9 ft (2.1–2.7 m)

pointed tip

concave trailing edge

broad base

robust body tapers to tail

broad, concave trailing edges

slight notch in middle

FLUKES

underside paler than upper side and sides and sometimes pinkish in color

body may appear wrinkled

MALE/FEMALE

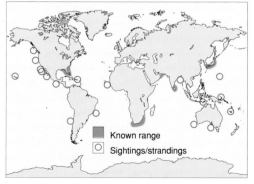

Known range

Sightings/strandings

DEEP TEMPERATE, SUBTROPICAL, AND TROPICAL WATERS OF
THE NORTHERN AND SOUTHERN HEMISPHERES

WHERE TO LOOK

Predominantly a deep-water species, possibly concentrated over the edge of the continental shelf (closer to shore than the Pygmy Sperm Whale). Appears to prefer warmer waters and seems to be especially common off the southern tip of Africa and in the Gulf of California (Sea of Cortez), Mexico, where it occurs particularly close to shore. Most records are from strandings, which are relatively common in some places, though these may simply represent areas of most research rather than a true picture of distribution. Lack of records of live animals may be due to inconspicuous behavior rather than rarity. Populations may be continuous around the world.

Birth wt 90–110 lb (40–50 kg)	Adult wt 300–605 lb (135–275 kg)	Diet

Family PHYSETERIDAE	Species *Physeter macrocephalus*	Habitat 〰〰〰 ▰〰〰

SPERM WHALE

The Sperm Whale is one of the easiest whales to identify at sea, even though it rarely shows much of itself at the surface. At a distance, its angled, bushy blow is usually sufficient for identification. At close range, its huge, squarish head (typically measuring a third of the body length) and wrinkly, prunelike skin are unmistakable. There are marked differences between the sexes: males average 49¼–59 ft (15–18 m) and females only 36–39½ ft (11–12 m). There are two main groupings: "bachelor schools" (young, sexually inactive males) and "breeding schools" (females with young of both sexes). These typically contain 20 to 25 animals although, in a few exceptional cases, hundreds or even thousands have been reported together. Older males tend to be solitary or live in small groups of up to 6 animals, joining the breeding schools for a few hours at a time during the breeding season. The Sperm Whale was one of the most heavily exploited of all the world's whales, although it is still relatively abundant.
• OTHER NAMES Great Sperm Whale, Cachalot, *P. catodon.*

slightly raised, slitlike blowhole on left side near • front of head

head proportionately larger in males • than females

small, inconspicuous • eyes

older males may be badly scarred, • especially around head

• blunt snout may extend up to 5 ft (1.5 m) beyond tip of lower jaw

• lower jaw barely visible when mouth closed

• large, squarish head, sometimes with gray or off-white areas

• short, stubby flippers

BEHAVIOR
Can remain submerged for over 2 hours, but typical dive time is less than 45 minutes. Interval between dives may be up to an hour, but usually 5 to 15 minutes. Breathes at regular 12- to 20-second intervals. Whalers' rule of thumb generally works very well: for every foot (30 cm) of its length, the Sperm Whale will breathe once at the surface and spend about 1 minute underwater during the next dive. Often surfaces near the same place. After a long dive, first exhalation is often strong and loud. When at the surface, usually remains almost motionless, but may swim leisurely. Capable of high speed if alarmed. Frequently breaches and lobtails. Sometimes strands.

• thick, conical teeth grow to 8 in (20 cm) long and may weigh over 2¼ lb (1 kg); females have fewer, smaller teeth

TEETH $\frac{0}{36-50}$

Group size 1–50 (1–150), hundreds may travel together	Hump position Far behind center

| Status Locally common | Population Unknown | Threats 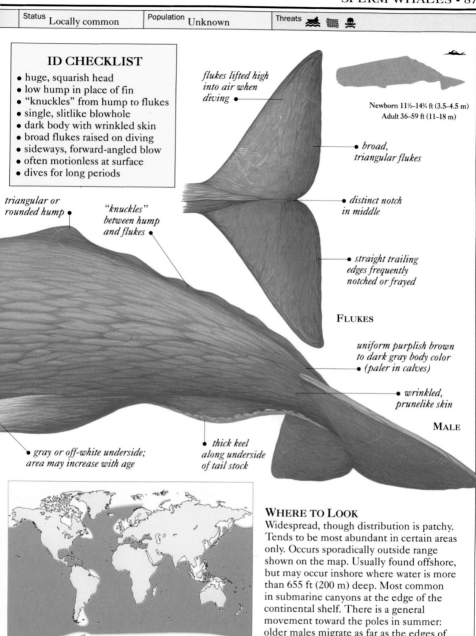 |

ID CHECKLIST

- huge, squarish head
- low hump in place of fin
- "knuckles" from hump to flukes
- single, slitlike blowhole
- dark body with wrinkled skin
- broad flukes raised on diving
- sideways, forward-angled blow
- often motionless at surface
- dives for long periods

flukes lifted high into air when diving

Newborn 11½–14¾ ft (3.5–4.5 m)
Adult 36–59 ft (11–18 m)

broad, triangular flukes

triangular or rounded hump

"knuckles" between hump and flukes

distinct notch in middle

straight trailing edges frequently notched or frayed

FLUKES

uniform purplish brown to dark gray body color (paler in calves)

wrinkled, prunelike skin

MALE

gray or off-white underside; area may increase with age

thick keel along underside of tail stock

WIDELY DISTRIBUTED IN DEEP WATERS WORLDWIDE, BOTH
OFFSHORE AND INSHORE

WHERE TO LOOK

Widespread, though distribution is patchy.
Tends to be most abundant in certain areas
only. Occurs sporadically outside range
shown on the map. Usually found offshore,
but may occur inshore where water is more
than 655 ft (200 m) deep. Most common
in submarine canyons at the edge of the
continental shelf. There is a general
movement toward the poles in summer:
older males migrate as far as the edges of
polar ice, but females and juveniles rarely
venture beyond 45° N or 42° S. Winter is
spent in temperate and tropical waters.
Some populations are resident year-round.

| Birth wt 1 ton | Adult wt 20–50 tons | Diet |

Family PHYSETERIDAE	Species *Physeter macrocephalus*	Habitat 〰〰 ▰〰

HEAD

The Sperm Whale's head is large because it contains a huge cavity called the spermaceti organ. This is believed to be used in buoyancy control and may also be used to focus sonar clicks. The organ contains a mass of weblike pipes filled with a yellowy wax. This wax can be cooled or heated, possibly by water sucked in through the blowhole, and thus shrinks and increases in density (helping the whale to sink) or expands and decreases in density (lifting the whale back toward the surface).

HEAD-ON VIEW

large head contains spermaceti organ

DIVING

After re-oxygenating, the whale throws its flukes and the rear third of its body high into the air, and then drops vertically toward the sea floor. Sperm Whales typically dive to depths of 985–1,965 ft (300–600 m), though some evidence suggests that they can dive to depths of at least 9,845 ft (3,000 m). Researchers often use hydrophones to detect the whales' echolocation clicking underwater. Most long, deep dives are made by older males. They breathe in deeply before descending but, on extreme dives, their lungs collapse and they rely on the vast store of oxygen in their muscles and blood; their heartbeat slows down, and the oxygen is sent only to parts of the body that need it most (primarily the heart and brain).

huge, square head

snout conceals lower jaw

arches back

drops vertically toward sea floor

dive speed reaches 39 in– 9¾ ft (1–3 m) per second

DIVE SEQUENCE
1. The whale lifts its head out of the water for a final breath. Only two-thirds of the body length can be seen during the blow.

2. The body straightens out and, after gently arching its back, the whale may disappear from sight, dropping just below the surface.

Group size 1–50 (1–150), hundreds may travel together	Hump position Far behind center

Status Locally common	Population Unknown	Threats

SKULL

The Sperm Whale has an unusually long, flat upper jaw, designed to support the bulk of its head. Its Y-shaped lower jaw is extremely long and narrow, with rows of rounded conical teeth.

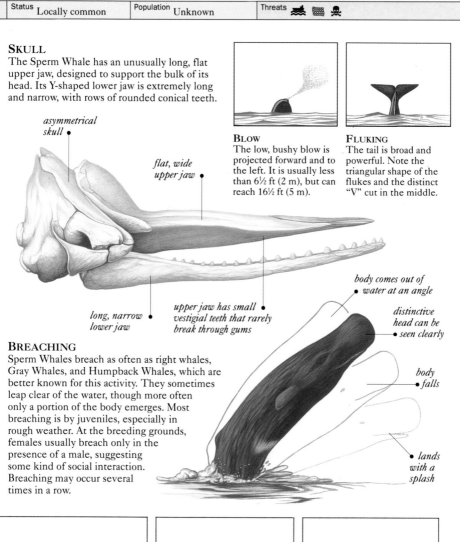

asymmetrical skull •

flat, wide upper jaw •

BLOW
The low, bushy blow is projected forward and to the left. It is usually less than 6½ ft (2 m), but can reach 16½ ft (5 m).

FLUKING
The tail is broad and powerful. Note the triangular shape of the flukes and the distinct "V" cut in the middle.

• *long, narrow lower jaw*

upper jaw has small vestigial teeth that rarely break through gums •

body comes out of water at an angle •

distinctive head can be seen clearly •

body falls •

BREACHING

Sperm Whales breach as often as right whales, Gray Whales, and Humpback Whales, which are better known for this activity. They sometimes leap clear of the water, though more often only a portion of the body emerges. Most breaching is by juveniles, especially in rough weather. At the breeding grounds, females usually breach only in the presence of a male, suggesting some kind of social interaction. Breaching may occur several times in a row.

lands with a splash •

3. After accelerating forward, the whale reappears (part of its back, hump, and head are visible) and begins to arch its back again.

4. With the back high out of water, the rounded hump and "knuckles" along the upper side are clearly visible.

5. The flukes and rear third of the body are thrown high into the air and drop vertically, barely creating a ripple.

Birth wt 1 ton	Adult wt 20–50 tons	Diet

NARWHAL AND BELUGA

T HIS IS A SMALL FAMILY, consisting of 2 medium-sized, gregarious whales living in the cold waters of the subarctic and Arctic. They are locally common, and sometimes travel or feed together, but tend to inhabit fairly remote and inaccessible areas. As they are unusual in both appearance and behavior, they have given rise to much outlandish folklore over the centuries. Many taxonomists include the Irrawaddy Dolphin (p.222) in this family. In some ways, it is the tropical equivalent of the Beluga: the two species are broadly similar in their appearance; there are some anatomical similarities, particularly in their skulls; and they are the only cetaceans that can dramatically change their facial expressions. Like the Beluga and the Narwhal, the Irrawaddy Dolphin has a very flexible neck, because in many animals all the cervical vertebrae, or nearly all of them, are unfused.

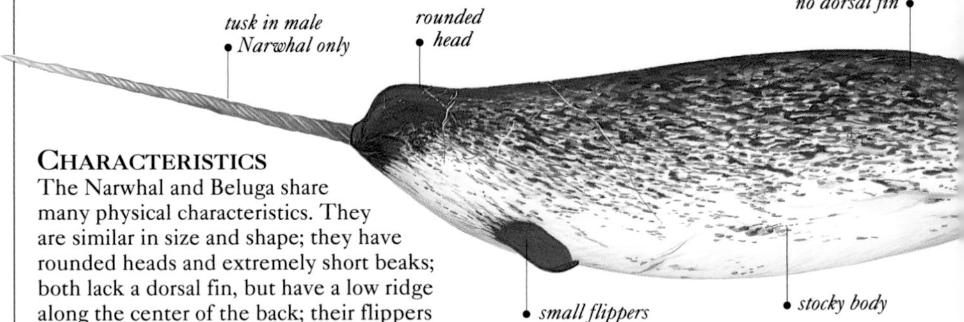

tusk in male • Narwhal only

rounded • head

no dorsal fin •

CHARACTERISTICS
The Narwhal and Beluga share many physical characteristics. They are similar in size and shape; they have rounded heads and extremely short beaks; both lack a dorsal fin, but have a low ridge along the center of the back; their flippers are small and rounded and have a tendency to curl up at the tips; their flukes have a distinct notch in the middle; and both species have thick layers of blubber for insulation against the cold, arctic seas. The calves of both the Narwhal and Beluga are darker in color than in the adults.

small flippers •

stocky body •

NARWHAL
There is a considerable difference between the sexes in both the Narwhal and the Beluga: in both species, males are larger than females. The Narwhal's tusk, usually present only in males, is the most obvious example.

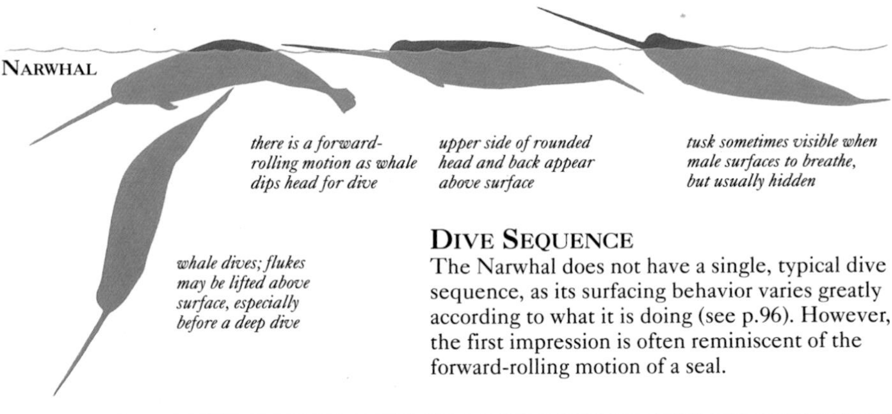

NARWHAL

there is a forward-rolling motion as whale dips head for dive

upper side of rounded head and back appear above surface

tusk sometimes visible when male surfaces to breathe, but usually hidden

whale dives; flukes may be lifted above surface, especially before a deep dive

DIVE SEQUENCE
The Narwhal does not have a single, typical dive sequence, as its surfacing behavior varies greatly according to what it is doing (see p.96). However, the first impression is often reminiscent of the forward-rolling motion of a seal.

BREEDING BELUGAS
Belugas are very social animals. In the summer hundreds, or even thousands, gather to give birth to their calves and molt their skin in shallow bays and estuaries of the subarctic and Arctic.

SPECIES IDENTIFICATION

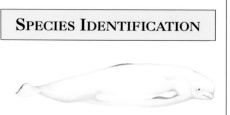

BELUGA (p.92) *Unmistakable whale with a stocky white or yellowish body, rounded head, short beak, and no dorsal fin.*

NARWHAL (p.96) *Beautifully mottled whale living farther north than almost any other cetacean; the male has a long, spiraling tusk.*

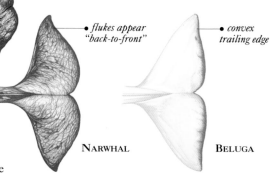

• *flukes appear "back-to-front"*

• *convex trailing edge*

distinct notch in flukes

NARWHAL

BELUGA

FLUKES
The Narwhal and Beluga have unusually shaped flukes, with convex trailing edges. They seem to face backward in the Narwhal and, although this appearance is not as pronounced in the Beluga, the trailing edges of its flukes do become more convex with age.

BELUGA

upper side of rounded forehead breaks the surface first

part of rounded back appears briefly

whale rises toward the surface at a shallow angle

whale dives; flukes normally remain hidden below the surface

DIVE SEQUENCE
It is often difficult to get a good view of a Beluga: it is usually visible for an average of only a few seconds when it surfaces to breathe, and only a small part of its body is exposed; the flukes normally remain below the surface. The lasting impression is of a gently undulating motion.

Family MONODONTIDAE	Species *Delphinapterus leucas*	Habitat 🐋 (🌊)

BELUGA

Adult Belugas are unlikely to be confused with any other cetacean because their uniform light coloration is so distinctive. However, they can be surprisingly difficult to pick out among whitecaps or floating ice: look for a white arc that appears, grows, shrinks, then disappears. The Beluga is one of the most vocal of the toothed whales, with a great repertoire of trills, moos, clicks, squeaks, and twitters, which can be heard above and below the surface. It may also have the most versatile and sophisticated sonar system of any cetacean. Well adapted to living close to shore, it can maneuver in very shallow water and swim in depths barely covering its body; if stranded and unmolested, it can often survive until the next tide.
• **OTHER NAMES** Belukha, Sea Canary, White Whale.

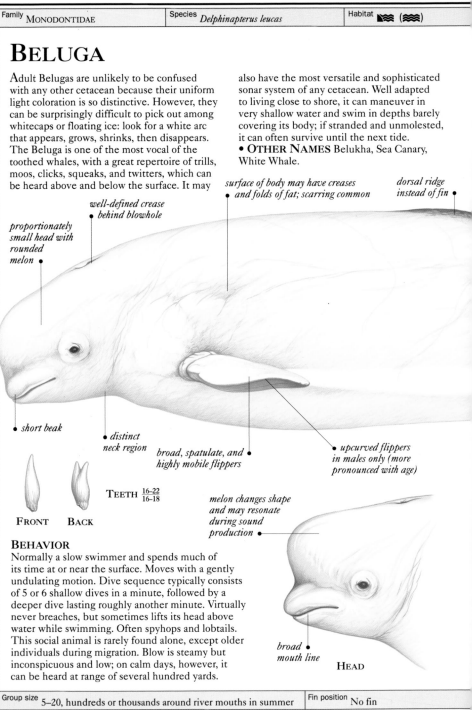

surface of body may have creases and folds of fat; scarring common

dorsal ridge instead of fin •

• well-defined crease behind blowhole

proportionately small head with rounded melon •

• short beak

• distinct neck region

broad, spatulate, and highly mobile flippers

• upcurved flippers in males only (more pronounced with age)

TEETH $\frac{16-22}{16-18}$

FRONT   BACK

melon changes shape and may resonate during sound production •

BEHAVIOR
Normally a slow swimmer and spends much of its time at or near the surface. Moves with a gently undulating motion. Dive sequence typically consists of 5 or 6 shallow dives in a minute, followed by a deeper dive lasting roughly another minute. Virtually never breaches, but sometimes lifts its head above water while swimming. Often spyhops and lobtails. This social animal is rarely found alone, except older individuals during migration. Blow is steamy but inconspicuous and low; on calm days, however, it can be heard at range of several hundred yards.

broad • mouth line

HEAD

Group size 5–20, hundreds or thousands around river mouths in summer	Fin position No fin

Status Locally common	Population 50,000–70,000	Threats

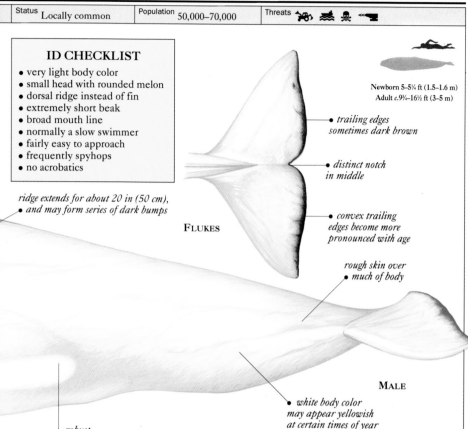

ID CHECKLIST

- very light body color
- small head with rounded melon
- dorsal ridge instead of fin
- extremely short beak
- broad mouth line
- normally a slow swimmer
- fairly easy to approach
- frequently spyhops
- no acrobatics

Newborn 5–5¼ ft (1.5–1.6 m)
Adult *c*.9¾–16½ ft (3–5 m)

• *trailing edges sometimes dark brown*

• *distinct notch in middle*

ridge extends for about 20 in (50 cm), • and may form series of dark bumps

FLUKES

• *convex trailing edges become more pronounced with age*

rough skin over • much of body

MALE

• *white body color may appear yellowish at certain times of year*

• *robust body shape*

WHERE TO LOOK

Wide but discontinuous circumpolar distribution from the subarctic to high arctic. Found off the coasts of Scandinavia, Greenland, Svalbard, former Soviet Union, and North America. Seasonal distribution is directly related to ice conditions, but most populations do not make extensive migrations; longest is by those that winter in the Bering Sea and summer in the Mackenzie River, Canada. In summer, some populations may swim 620 miles (1,000 km) or more up river. Other populations do not migrate at all, such as the residents of St. Lawrence River, Canada. Spends summer in shallow bays and estuaries. Winters in areas of loose pack ice, where wind and ocean currents keep cracks and breathing holes open.

CIRCUMPOLAR DISTRIBUTION IN SEASONALLY ICE-COVERED WATERS OF THE ARCTIC AND SUBARCTIC

Birth wt 175 lb (80 kg)	Adult wt 0.4–1.5 tons	Diet

Family MONODONTIDAE	Species *Delphinapterus leucas*	Habitat ◣▨ (▨)

BELUGA POPULATIONS

Five main Beluga populations are recognized: in the Bering, Chukchi, and Okhotsk Seas (25,000 to 30,000); high-arctic Canada and west Greenland (10,000 to 14,000); Hudson Bay and James Bay, Canada (9,000 to 12,000); Svalbard area (5,000 to 10,000); and Gulf of St. Lawrence, Canada (300 to 500). The St. Lawrence animals have such high concentrations of chemical contaminants in their bodies that they are treated as toxic waste when they die. Human disturbance, through oil exploration and hydroelectric plants on rivers used for calving, is of great concern. Belugas have been hunted by Arctic native people for hundreds of years, but overhunting by commercial operators in the 20th century has reduced their numbers. Body size varies considerably between the populations: the largest live off Greenland and in the Sea of Okhotsk and the smallest inhabit the White Sea and Hudson Bay.

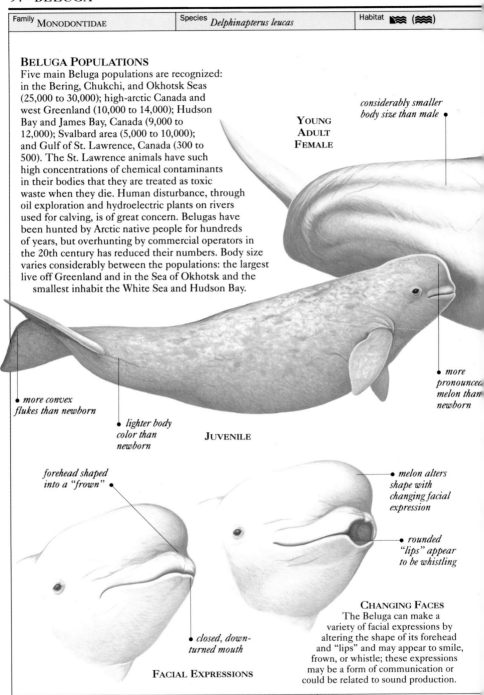

YOUNG ADULT FEMALE

considerably smaller body size than male •

• *more pronounced melon than newborn*

• *more convex flukes than newborn*

• *lighter body color than newborn*

JUVENILE

forehead shaped into a "frown" •

• *melon alters shape with changing facial expression*

• *rounded "lips" appear to be whistling*

• *closed, down-turned mouth*

CHANGING FACES

The Beluga can make a variety of facial expressions by altering the shape of its forehead and "lips" and may appear to smile, frown, or whistle; these expressions may be a form of communication or could be related to sound production.

FACIAL EXPRESSIONS

Group size 5–20, hundreds or thousands around river mouths in summer	Fin position No fin

Status		Population		Threats	
Locally common		50,000–70,000		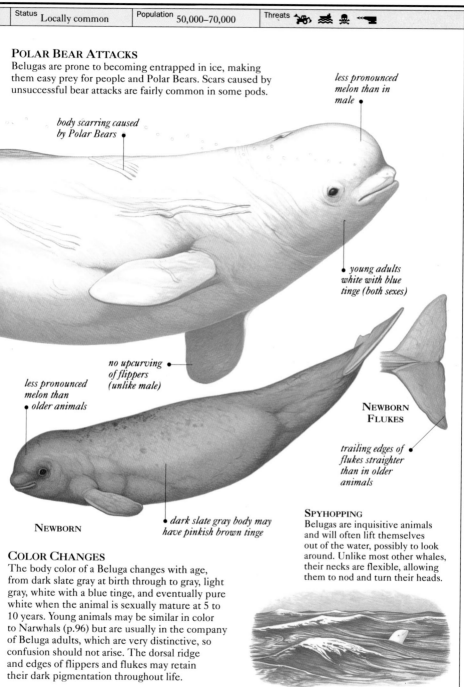	

POLAR BEAR ATTACKS

Belugas are prone to becoming entrapped in ice, making them easy prey for people and Polar Bears. Scars caused by unsuccessful bear attacks are fairly common in some pods.

less pronounced melon than in male

body scarring caused by Polar Bears

young adults white with blue tinge (both sexes)

no upcurving of flippers (unlike male)

less pronounced melon than older animals

NEWBORN FLUKES

trailing edges of flukes straighter than in older animals

NEWBORN

dark slate gray body may have pinkish brown tinge

COLOR CHANGES

The body color of a Beluga changes with age, from dark slate gray at birth through to gray, light gray, white with a blue tinge, and eventually pure white when the animal is sexually mature at 5 to 10 years. Young animals may be similar in color to Narwhals (p.96) but are usually in the company of Beluga adults, which are very distinctive, so confusion should not arise. The dorsal ridge and edges of flippers and flukes may retain their dark pigmentation throughout life.

SPYHOPPING

Belugas are inquisitive animals and will often lift themselves out of the water, possibly to look around. Unlike most other whales, their necks are flexible, allowing them to nod and turn their heads.

Birth wt	Adult wt	Diet
175 lb (80 kg)	0.4–1.5 tons	

Family MONODONTIDAE	Species *Monodon monoceros*	Habitat 🐋 (🌊)

NARWHAL

The male Narwhal is unique and unlikely to be confused with any other cetacean. Its long, spiraling tusk, which is actually a modified tooth, looks like a gnarled and twisted walking stick and, until early in the 17th century, was believed by many to have been the horn of the legendary unicorn. The role of the tusk baffled scientists for years. Among the theories were that it was used for spearfishing, grubbing for food, and drilling through ice; it is true that Narwhals are often trapped underwater by rapid freezing but, instead of using their tusks, they head-butt the necessary breathing holes. In fact, the tusk is probably used like the antlers of a deer, in fights over females and as a visual display of strength; roughly 1 in 3 tusks are broken, and the heads of most older males are covered with scars from fighting. The majority of females do not possess tusks, and older animals can be almost entirely white, so some confusion with Belugas (p.92) may occur.

• **OTHER NAME** Narwhale.

slight hump instead of dorsal fin •

flexible neck, with little • evidence of neck crease

proportionately • small head

bulbous • forehead

• slight hint of beak

• small, upturned mouth

tip brilliant white and • usually smoothly polished

• short flippers

TEETH $\frac{0-2}{0}$

brittle tusk hollow for most of its length •

BEHAVIOR

When feeding, moves erratically and spends little time at surface, usually diving for 7 to 20 minutes; on migration, swims fast and stays at or close to surface; when hunting, mills around or moves very slowly. All members of a group may surface and dive at the same time. May rest at surface for up to 10 minutes, with part of back or a flipper above water; in rough seas, tends to rest at depth. Tusk may be raised above surface. Spy-hopping, lobtailing, and flipper-slapping fairly common. Rarely breaches, but sometimes lunges at surface when swimming. Weak and usually inconspicuous blow.

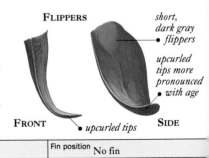

FLIPPERS

short, dark gray • flippers

upcurled tips more pronounced • with age

FRONT

• upcurled tips

SIDE

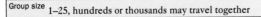

Group size 1–25, hundreds or thousands may travel together	Fin position No fin

Status Locally common	Population 25,000–45,000	Threats

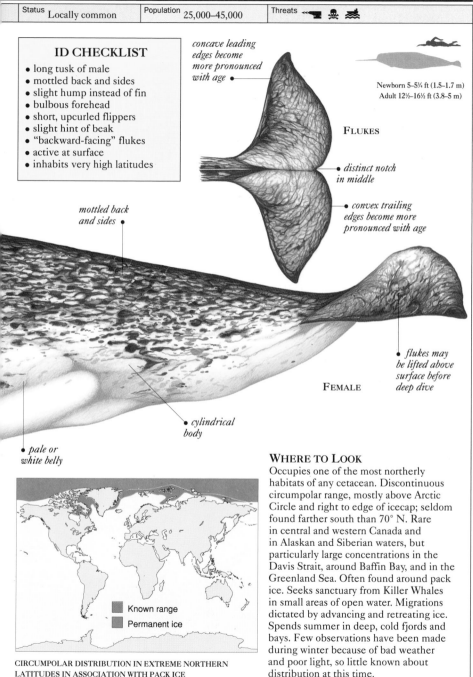

ID CHECKLIST

- long tusk of male
- mottled back and sides
- slight hump instead of fin
- bulbous forehead
- short, upcurled flippers
- slight hint of beak
- "backward-facing" flukes
- active at surface
- inhabits very high latitudes

Newborn 5–5¾ ft (1.5–1.7 m)
Adult 12½–16½ ft (3.8–5 m)

concave leading edges become more pronounced with age

FLUKES

distinct notch in middle

mottled back and sides

convex trailing edges become more pronounced with age

flukes may be lifted above surface before deep dive

FEMALE

cylindrical body

pale or white belly

WHERE TO LOOK

Occupies one of the most northerly habitats of any cetacean. Discontinuous circumpolar range, mostly above Arctic Circle and right to edge of icecap; seldom found farther south than 70° N. Rare in central and western Canada and in Alaskan and Siberian waters, but particularly large concentrations in the Davis Strait, around Baffin Bay, and in the Greenland Sea. Often found around pack ice. Seeks sanctuary from Killer Whales in small areas of open water. Migrations dictated by advancing and retreating ice. Spends summer in deep, cold fjords and bays. Few observations have been made during winter because of bad weather and poor light, so little known about distribution at this time.

Known range
Permanent ice

CIRCUMPOLAR DISTRIBUTION IN EXTREME NORTHERN LATITUDES IN ASSOCIATION WITH PACK ICE

Birth wt 175 lb (80 kg)	Adult wt 0.8–1.6 tons	Diet

Family MONODONTIDAE	Species *Monodon monoceros*	Habitat 🌊 (🌊)

PREDATION AND HUNTING

The Narwhal's predators are Killer Whales, Walruses, Polar Bears, and sharks. Its main enemy, however, is Man. It has been hunted by Inuit for centuries for its valuable tusk and its thick skin, which is traditionally eaten raw as a delicacy; its meat is used for dog food, and the blubber and fat for heating and lighting. Today, Canadian Inuit hunt with fast motor boats and use high-powered rifles – a wasteful method, as at least half of the whales sink or escape and die later from their wounds; despite an annual quota of a few hundred, the total kill is higher.

The Greenland Inuit have a similar quota, but hunt using kayaks and hand-held harpoons, avoiding most of the waste. Commercial whalers from the 17th century occasionally hunted Narwhal, but they usually bartered goods for the tusks; most were sold to China and Japan, where they still believed in unicorns and the potency of their horns in medicine.

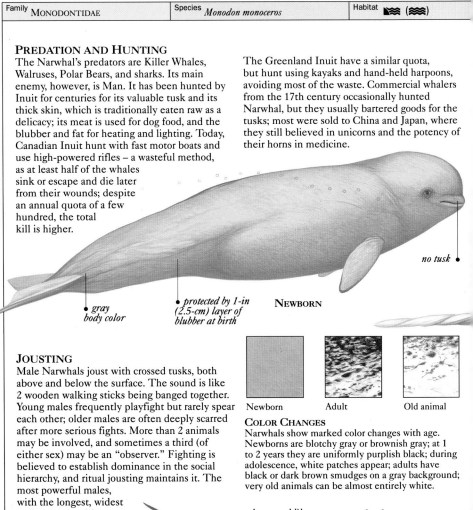

• *no tusk*

NEWBORN

• *gray body color*

• *protected by 1-in (2.5-cm) layer of blubber at birth*

JOUSTING

Male Narwhals joust with crossed tusks, both above and below the surface. The sound is like 2 wooden walking sticks being banged together. Young males frequently playfight but rarely spear each other; older males are often deeply scarred after more serious fights. More than 2 animals may be involved, and sometimes a third (of either sex) may be an "observer." Fighting is believed to establish dominance in the social hierarchy, and ritual jousting maintains it. The most powerful males, with the longest, widest tusks, may be able to mate with more females.

Newborn	Adult	Old animal

COLOR CHANGES

Narwhals show marked color changes with age. Newborns are blotchy gray or brownish gray; at 1 to 2 years they are uniformly purplish black; during adolescence, white patches appear; adults have black or dark brown smudges on a gray background; very old animals can be almost entirely white.

tusks crossed like swords •

scars on head • *from fighting*

JOUSTING MALES

Group size 1–25, hundreds or thousands may travel together	Fin position No fin

Status	Population	Threats
Locally common	25,000–45,000	

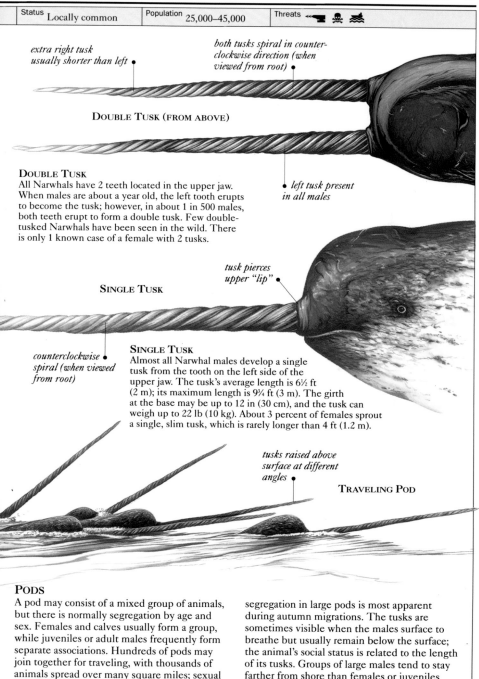

extra right tusk usually shorter than left

both tusks spiral in counter-clockwise direction (when viewed from root)

DOUBLE TUSK (FROM ABOVE)

DOUBLE TUSK
All Narwhals have 2 teeth located in the upper jaw. When males are about a year old, the left tooth erupts to become the tusk; however, in about 1 in 500 males, both teeth erupt to form a double tusk. Few double-tusked Narwhals have been seen in the wild. There is only 1 known case of a female with 2 tusks.

left tusk present in all males

tusk pierces upper "lip"

SINGLE TUSK

counterclockwise spiral (when viewed from root)

SINGLE TUSK
Almost all Narwhal males develop a single tusk from the tooth on the left side of the upper jaw. The tusk's average length is 6½ ft (2 m); its maximum length is 9¾ ft (3 m). The girth at the base may be up to 12 in (30 cm), and the tusk can weigh up to 22 lb (10 kg). About 3 percent of females sprout a single, slim tusk, which is rarely longer than 4 ft (1.2 m).

tusks raised above surface at different angles

TRAVELING POD

PODS
A pod may consist of a mixed group of animals, but there is normally segregation by age and sex. Females and calves usually form a group, while juveniles or adult males frequently form separate associations. Hundreds of pods may join together for traveling, with thousands of animals spread over many square miles; sexual segregation in large pods is most apparent during autumn migrations. The tusks are sometimes visible when the males surface to breathe but usually remain below the surface; the animal's social status is related to the length of its tusks. Groups of large males tend to stay farther from shore than females or juveniles.

Birth wt	Adult wt	Diet
175 lb (80 kg)	0.8–1.6 tons	

BEAKED WHALES

BEAKED WHALES are the least known of all cetaceans. Indeed, some have never been seen alive. Most are known from dead animals washed ashore and, in some cases, from a few brief encounters at sea. Some may be genuinely rare, or simply elusive, but the main problem is that they generally live in deep water far from land. They are small- to medium-sized whales and range in length from under 13¼ ft (4 m) to nearly 42¾ ft (13 m). To identify a beaked whale by color alone is unreliable, because of the tremendous variation between individuals and, in any case, little is known about the coloration of live animals. There are currently 20 known species, but there could be others that have yet to be discovered.

slender beak •

• no real distinction between head and beak

small dorsal fin set far back on body •

V-shaped • throat grooves

CHARACTERISTICS
The most remarkable feature of beaked whales is the teeth of the males: their shape and position are often crucial for a positive identification. Most males have only 2 teeth (*Mesoplodon*, *Ziphius*, and *Hyperoodon* species) or 4 teeth (*Berardius* species) in the lower jaw and none in the upper jaw. Most females have no teeth at all. Gray's and Shepherd's Beaked Whales are the only exceptions, with extra rows of tiny teeth erupting in both sexes.

• small flippers

• spindle-shaped body

conditions in ocean trenches were probably similar •

evolving populations were effectively isolated from one another •

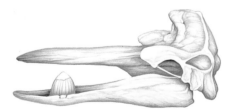

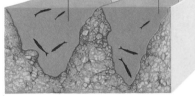

SKULL
The skulls of beaked whales (except *Berardius* species) are asymmetrical. The dramatically reduced number of teeth is typical of those species that eat mainly squid, although some members of the family also eat deep-sea fish.

LIVING APART
Most beaked whale sightings are concentrated along deep ocean trenches. The whales in different trenches have evolved in isolation but have adapted to similar conditions; this may explain their physical similarities.

BLAINVILLE'S BEAKED WHALE
This is one of the few underwater photographs taken of a beaked whale at sea; note the strongly arched lower jaw, flattened forehead, slender beak, and light-colored blotches.

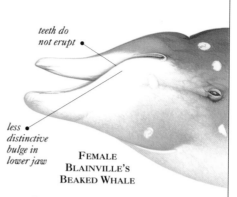

teeth do not erupt

less distinctive bulge in lower jaw

FEMALE BLAINVILLE'S BEAKED WHALE

distinctive pair of teeth in middle of lower jaw

substantial bulge in lower jaw

MALE BLAINVILLE'S BEAKED WHALE

flukes usually have no notch in middle

HECTOR'S BEAKED WHALE
Hector's Beaked Whale shows several physical characteristics common to all beaked whales. The V-shaped throat grooves are unique to this family and conspicuous on examination of dead animals. The scratches and scars, caused by teeth during fights, are typical of most males.

TEETH
The teeth of female and young beaked whales are normally absent or invisible, making them almost impossible to identify at sea; however, the males have tusklike teeth that, with experience and optimum conditions, can be used for identification.

NORTHERN BOTTLENOSE WHALE

flukes may be lifted above surface in preparation for dive

after blowing, whale cruises before diving (lying notably high in the water)

blunt forehead and back (as far as dorsal fin) usually visible as whale surfaces

dives are deep, but whale typically does not travel much horizontal distance while submerged

DIVE SEQUENCE
Beaked whales are rarely seen at sea. They are shy and spend long periods underwater, and they are relatively inconspicuous at the surface. They also usually live alone or in small groups far from land. As a result, they are hard to identify and are easily confused with other species.

SPECIES IDENTIFICATION

LESSER BEAKED WHALE (p.136) *Smallest member of the family, with little body scarring and no visible teeth; officially named in 1991.*

GRAY'S BEAKED WHALE (p.126) *Appears to be more social and active at the surface than other beaked whales; long, slender beak is usually white.*

HECTOR'S BEAKED WHALE (p.128) *Small whale; male has triangular teeth near the tip of the lower jaw; known mainly from skeletons and skulls.*

TRUE'S BEAKED WHALE (p.132) *Yet to be positively identified at sea; male has 2 small teeth at the tip of its moderate-sized beak.*

ANDREWS' BEAKED WHALE (p.116) *Very similar to Hubbs' Beaked Whale; male has massive teeth in the middle of a strongly arched mouth line.*

BLAINVILLE'S BEAKED WHALE (p.120) *Flat forehead and spotty body; male has a strongly arched lower jaw with a pair of massive teeth.*

SOWERBY'S BEAKED WHALE (p.114) *Bulge in front of the blowhole and limited scarring on body and head; there have been few sightings at sea.*

HUBBS' BEAKED WHALE (p.118) *Distinctive raised white "cap" on head, 2 massive teeth, stocky white beak, and a tangle of scars on its body.*

GERVAIS' BEAKED WHALE (p.122) *Especially hard to identify at sea, but the position of the teeth and the narrow beak of the male may be distinctive.*

STEJNEGER'S BEAKED WHALE (p.138) *Dark, gently sloping forehead; male has 2 huge, broad, flat teeth and strongly arched mouth line.*

GINKGO-TOOTHED BEAKED WHALE (p.124) *Male's teeth are shaped like Ginkgo Tree leaves; little body scarring; white spots around navel of male.*

UNIDENTIFIED BEAKED WHALE (p.112) *Known only from about 30 sightings at sea; sexes appear to differ significantly in color.*

SPECIES IDENTIFICATION

STRAP-TOOTHED WHALE (p.130) *Strangest-looking member of the family; the teeth of older males curl over the upper jaw, preventing the mouth from opening properly.*

SHEPHERD'S BEAKED WHALE (p.140) *One of the least known cetaceans, but may be identifiable at sea by its long, narrow beak, steep forehead, and diagonal striping on the sides.*

CUVIER'S BEAKED WHALE (p.142) *Widespread and relatively abundant, but rarely seen; head shaped like a goose's beak and has long and circular scars over its body.*

SOUTHERN BOTTLENOSE WHALE (p.110) *Huge, bulbous forehead, distinct dolphinlike beak, and robust, cylindrical body; male has a pair of teeth at the tip of the lower jaw.*

LONGMAN'S BEAKED WHALE (p.134) *Known from only 2 weathered skulls, though there have been several possible sightings; this illustration is an artist's impression.*

NORTHERN BOTTLENOSE WHALE (p.108) *Northern equivalent of the Southern Bottlenose Whale and similar in appearance, but typically larger.*

ARNOUX'S BEAKED WHALE (p.104) *Closely resembles Baird's Beaked Whale but is geographically separate; has extensively scarred body, and teeth erupt in both sexes.*

BAIRD'S BEAKED WHALE (p.106) *Largest member of the family; closely resembles Arnoux's Beaked Whale but is better known.*

Family ZIPHIIDAE	Species *Berardius arnuxii*	Habitat 〜〜〜

ARNOUX'S BEAKED WHALE

Arnoux's Beaked Whale is very poorly known and appears to be relatively rare. It resembles Baird's Beaked Whale (p.106) so closely that some believe it to be the same species. However, the two appear to be geographically isolated, and Arnoux's is probably smaller; observations of live animals suggest a length of up to 39½ ft (12 m), but all dead specimens examined have been considerably shorter. Confusion is most likely with the Southern Bottlenose Whale (p.110), from which it is almost indistinguishable at sea; however, look for the longer beak and less bulbous forehead of Arnoux's. It is unusual among beaked whales as the teeth erupt in both males and females; in older animals, they may be worn down to the gums. Although dark in color, the body may appear pale brown or even orange at sea; this is caused by a covering of microscopic algae, called diatoms, over the body.
• OTHER NAMES Southern Four-toothed Whale, Southern Beaked Whale, New Zealand Beaked Whale, Southern Giant Bottlenose Whale, Southern Porpoise Whale.

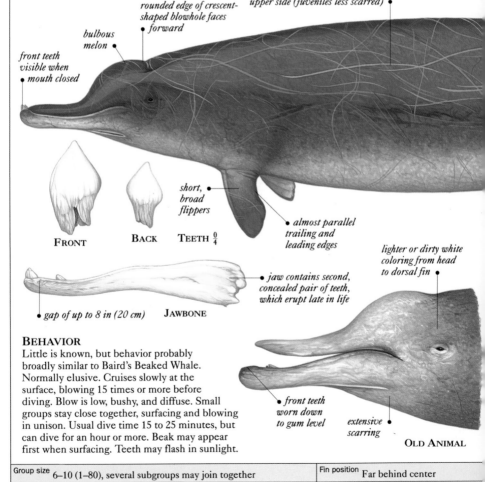

heavy white scarring, especially on upper side (juveniles less scarred) •

rounded edge of crescent-shaped blowhole faces • *forward*

bulbous melon •

front teeth visible when • *mouth closed*

FRONT   **BACK**   **TEETH** $\frac{0}{4}$

short, broad flippers •

• *almost parallel trailing and leading edges*

lighter or dirty white coloring from head to dorsal fin •

• *gap of up to 8 in (20 cm)*   **JAWBONE**

• *jaw contains second, concealed pair of teeth, which erupt late in life*

BEHAVIOR
Little is known, but behavior probably broadly similar to Baird's Beaked Whale. Normally elusive. Cruises slowly at the surface, blowing 15 times or more before diving. Blow is low, bushy, and diffuse. Small groups stay close together, surfacing and blowing in unison. Usual dive time 15 to 25 minutes, but can dive for an hour or more. Beak may appear first when surfacing. Teeth may flash in sunlight.

• *front teeth worn down to gum level*

extensive • *scarring*

OLD ANIMAL

Group size 6–10 (1–80), several subgroups may join together	Fin position Far behind center

Status Unknown	Population Unknown	Threats Unknown

ID CHECKLIST

- extensively scarred body
- front teeth visible
- pronounced, dolphinlike beak
- bulbous melon
- spindle-shaped body
- small fin
- slow swimmer
- small, tightly packed groups
- often difficult to approach

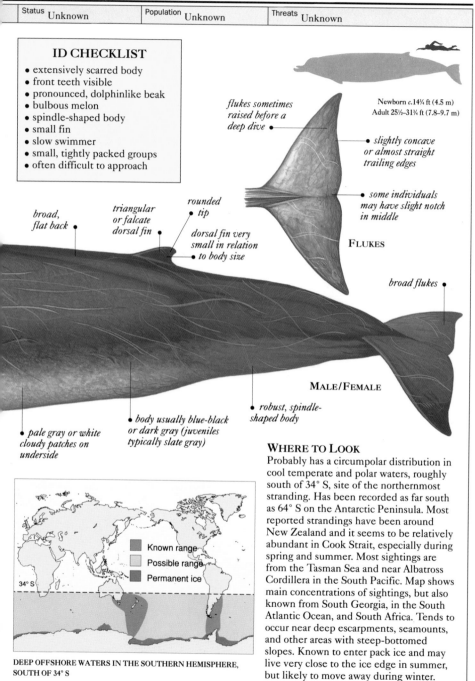

Newborn *c.*14¾ ft (4.5 m)
Adult 25½–31¾ ft (7.8–9.7 m)

flukes sometimes raised before a deep dive •

• *slightly concave or almost straight trailing edges*

• *some individuals may have slight notch in middle*

FLUKES

broad, flat back •

triangular or falcate dorsal fin

rounded • *tip*

dorsal fin very small in relation • *to body size*

broad flukes •

MALE/FEMALE

• *robust, spindle-shaped body*

• *pale gray or white cloudy patches on underside*

• *body usually blue-black or dark gray (juveniles typically slate gray)*

WHERE TO LOOK

Probably has a circumpolar distribution in cool temperate and polar waters, roughly south of 34° S, site of the northernmost stranding. Has been recorded as far south as 64° S on the Antarctic Peninsula. Most reported strandings have been around New Zealand and it seems to be relatively abundant in Cook Strait, especially during spring and summer. Most sightings are from the Tasman Sea and near Albatross Cordillera in the South Pacific. Map shows main concentrations of sightings, but also known from South Georgia, in the South Atlantic Ocean, and South Africa. Tends to occur near deep escarpments, seamounts, and other areas with steep-bottomed slopes. Known to enter pack ice and may live very close to the ice edge in summer, but likely to move away during winter.

34° S

Known range
Possible range
Permanent ice

DEEP OFFSHORE WATERS IN THE SOUTHERN HEMISPHERE, SOUTH OF 34° S

Birth wt Unknown	Adult wt 7–10 tons	Diet

| Family ZIPHIIDAE | Species *Berardius bairdii* | Habitat 〰 |

BAIRD'S BEAKED WHALE

Baird's is probably the largest of all the beaked whales. Its external appearance is so similar to Arnoux's Beaked Whale (p.104) that some believe it to be the same species. However, the two appear to be geographically isolated, and Baird's is probably slightly larger. Baird's is by far the better known. As with Arnoux's, the teeth erupt in both males and females; in older animals, they may be worn down to gum level. The front pair are especially distinctive and, in bright sunlight, often appear brilliant white against the dark body and surrounding sea. Confusion is possible with the form of Southern Bottlenose Whale found in the eastern tropical Pacific (p.111), though there may be no overlap in range. Juveniles may be confused with smaller beaked whale species but are likely to be in the company of larger adults. Small numbers of Baird's Beaked Whales have been hunted off the Boso Peninsula, Japan, for several hundred years; nowadays, 40 to 60 are taken annually under a government quota system.

• **OTHER NAMES** Northern Giant Bottlenose Whale, North Pacific Bottlenose Whale, Giant Four-toothed Whale, Northern Four-toothed Whale, North Pacific Four-toothed Whale.

upper side may appear much lighter if heavily scarred, especially in males

bulging forehead broader and more bulbous in males

indentation at blowhole

front teeth exposed when mouth closed

lower jaw protrudes beyond upper jaw

small, slightly rounded flippers

variable whitish spots and blotches on underside

flippers far forward on body

TEETH $\frac{0}{4}$

FRONT　**BACK**

exposed front teeth often heavily infested with whale lice and stalked barnacles

BEHAVIOR
May be wary of boats where hunted, but more approachable elsewhere. Low, bushy blow is sometimes visible. Forehead and beak often break the surface when animal rises to breathe. Blowhole normally disappears before dorsal fin emerges. Whole pod stays in tight formation, surfacing and blowing in unison. Typically visible for less than 5 minutes. Deep dives usually last 25 to 35 minutes. Spyhopping, lobtailing, logging, and (rarely) breaching observed.

second pair of teeth, concealed inside mouth, erupt late in life

JAWBONE

| Group size 3–30 (1–50), larger groups may split for short periods | Fin position Far behind center |

Status Locally common	Population Unknown	Threats Unknown

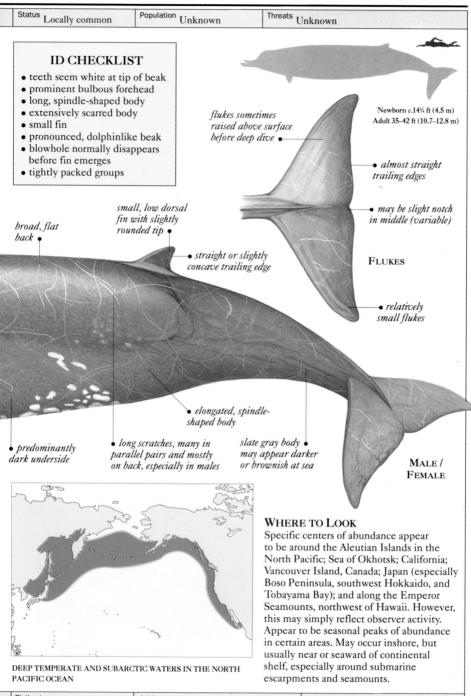

ID CHECKLIST

- teeth seem white at tip of beak
- prominent bulbous forehead
- long, spindle-shaped body
- extensively scarred body
- small fin
- pronounced, dolphinlike beak
- blowhole normally disappears before fin emerges
- tightly packed groups

flukes sometimes raised above surface before deep dive •

Newborn *c.*14¾ ft (4.5 m)
Adult 35–42 ft (10.7–12.8 m)

• almost straight trailing edges

• may be slight notch in middle (variable)

FLUKES

small, low dorsal fin with slightly rounded tip •

broad, flat back •

• straight or slightly concave trailing edge

• relatively small flukes

• elongated, spindle-shaped body

• predominantly dark underside

• long scratches, many in parallel pairs and mostly on back, especially in males

slate gray body • may appear darker or brownish at sea

MALE / FEMALE

DEEP TEMPERATE AND SUBARCTIC WATERS IN THE NORTH PACIFIC OCEAN

WHERE TO LOOK

Specific centers of abundance appear to be around the Aleutian Islands in the North Pacific; Sea of Okhotsk; California; Vancouver Island, Canada; Japan (especially Boso Peninsula, southwest Hokkaido, and Tobayama Bay); and along the Emperor Seamounts, northwest of Hawaii. However, this may simply reflect observer activity. Appear to be seasonal peaks of abundance in certain areas. May occur inshore, but usually near or seaward of continental shelf, especially around submarine escarpments and seamounts.

Birth wt Unknown	Adult wt *c.*11–15 tons	Diet

Family ZIPHIIDAE	Species *Hyperoodon ampullatus*	Habitat 〰️

NORTHERN BOTTLENOSE WHALE

The Northern Bottlenose Whale is a curious animal: it will approach stationary boats and seems to be attracted by strange noises, such as those made by ships' generators. This, combined with its habit of staying with wounded companions, made it especially vulnerable to whalers: tens of thousands were killed, mainly from 1850–1973. The species has been protected since 1977. Its most distinctive feature is the bulbous forehead; this is more pronounced in older animals and most distinct in adult males. There are usually 2 teeth; these erupt only in males and remain below the gums in females. Some males may have 4 teeth or none at all and there may be many toothpicklike vestigial teeth in both jaws of both sexes. Confusion may occur with pilot whales (pp.148–151), but the color, dorsal fin, and beak of the Northern Bottlenose are all distinctive. Minke Whales (p.56) have a similar fin but different head shape. Sowerby's (p.114) and Cuvier's Beaked Whales (p.142) are also similar but have less bulbous foreheads.
• **OTHER NAMES** North Atlantic Bottlenosed Whale, Flathead, Bottlehead, Steephead.

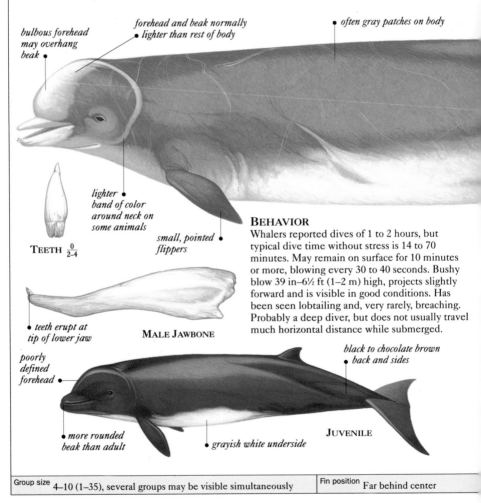

bulbous forehead may overhang beak •

forehead and beak normally • lighter than rest of body

• often gray patches on body

lighter • band of color around neck on some animals

TEETH $\frac{0}{2-4}$

small, pointed • flippers

• teeth erupt at tip of lower jaw

MALE JAWBONE

BEHAVIOR
Whalers reported dives of 1 to 2 hours, but typical dive time without stress is 14 to 70 minutes. May remain on surface for 10 minutes or more, blowing every 30 to 40 seconds. Bushy blow 39 in–6½ ft (1–2 m) high, projects slightly forward and is visible in good conditions. Has been seen lobtailing and, very rarely, breaching. Probably a deep diver, but does not usually travel much horizontal distance while submerged.

poorly defined forehead •

black to chocolate brown • back and sides

• more rounded beak than adult

• grayish white underside

JUVENILE

Group size 4–10 (1–35), several groups may be visible simultaneously	Fin position Far behind center

Status Unknown	Population Unknown	Threats 〰️ 🦑

ID CHECKLIST

- fin two-thirds of way down back
- bulbous forehead
- small, dolphinlike beak
- robust, cylindrical body
- no notch in flukes
- dark gray to brown color
- visible, bushy blow
- usually in small groups
- may be curious

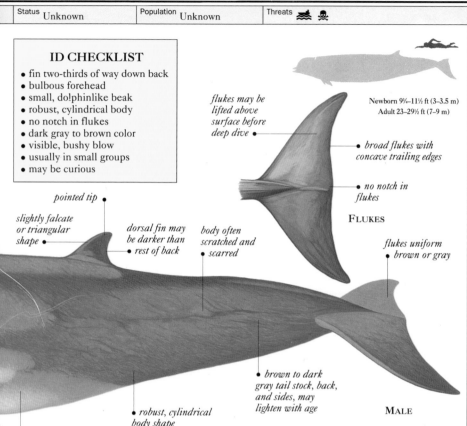

Newborn 9¾–11½ ft (3–3.5 m)
Adult 23–29½ ft (7–9 m)

flukes may be lifted above surface before deep dive •

• broad flukes with concave trailing edges

• no notch in flukes

FLUKES

pointed tip •

slightly falcate or triangular shape •

dorsal fin may be darker than • rest of back

body often scratched and • scarred

flukes uniform • brown or gray

• brown to dark gray tail stock, back, and sides, may lighten with age

MALE

• robust, cylindrical body shape

• creamy brown or pale gray underside

WHERE TO LOOK

Appear to be certain pockets of abundance: around "the Gully," north of Sable Island, Nova Scotia, Canada; in the Arctic Ocean, between Iceland and Jan Mayen and southwest of Svalbard; and in Davis Strait, off northern Labrador, Canada, especially around the entrance to Hudson Strait and Frobisher Bay. Less common in extreme southern part of range. In east of range probably moves north in spring and south in autumn; in the west, at least some animals believed to overwinter at higher latitudes. May also be some inshore–offshore movements. Most common beyond the continental shelf and over submarine canyons in deep water. Sometimes travels several miles into broken ice fields, but more common in open water. Known to strand.

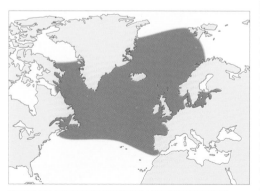

NORTH ATLANTIC OCEAN, NORMALLY IN WATER DEEPER THAN 3,280 FT (1,000 M)

Birth wt Unknown	Adult wt 5.8–7.5 tons	Diet ⤙ (▨ ★)

Family ZIPHIIDAE	Species *Hyperoodon planifrons*	Habitat 〰〰

SOUTHERN BOTTLENOSE WHALE

The Southern Bottlenose Whale is poorly known and rarely observed at sea. It lives far from shipping lanes and has never been heavily exploited, so it has not been as well studied as its northern counterpart (p.108). It has an extremely bulbous forehead, which is more pronounced in older animals and most distinct in adult males; in old males, the front of the forehead is almost vertical and flat. There are generally 2 teeth, which remain beneath the gums in females but erupt in males. However, some males may have 4 teeth or none at all. Both sexes may have vestigial, toothpicklike teeth in both jaws. The Southern Bottlenose Whale may be confused with the Minke Whale (p.56), Arnoux's Beaked Whale (p.104), or the Long-finned Pilot Whale (p.150), with which it sometimes associates.

• **OTHER NAMES** Antarctic Bottlenosed Whale, Flathead.

• *teeth erupt at tip of lower jaw*

MALE JAWBONE

bulbous forehead and small beak lighter • than rest of body

wide blowhole located • in indentation

• *indent above upper jaw*

small, tapering flippers •

• *pointed tips*

• *body frequently covered with scratches and scars*

TEETH $\frac{0}{2}$

BEHAVIOR

Few reports of swimming near boats, but this may be due to lack of observation rather than shyness. After long dive, may remain on surface for 10 minutes or more, blowing every 30 to 40 seconds. Bushy blow 39 in–6½ ft (1–2 m) high, projects slightly forward and visible in good conditions. Can stay underwater for at least an hour, but typical dive time is shorter. When swimming fast, especially under duress, may raise head clear of water on surfacing. Probably a deep diver, though does not tend to travel much horizontal distance while submerged.

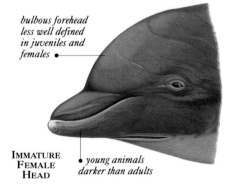

bulbous forehead less well defined in juveniles and females •

IMMATURE FEMALE HEAD

• *young animals darker than adults*

Group size 1–25, but fewer than 10 more common in Antarctic	Fin position Far behind center

Status Unknown	Population Unknown	Threats Unknown

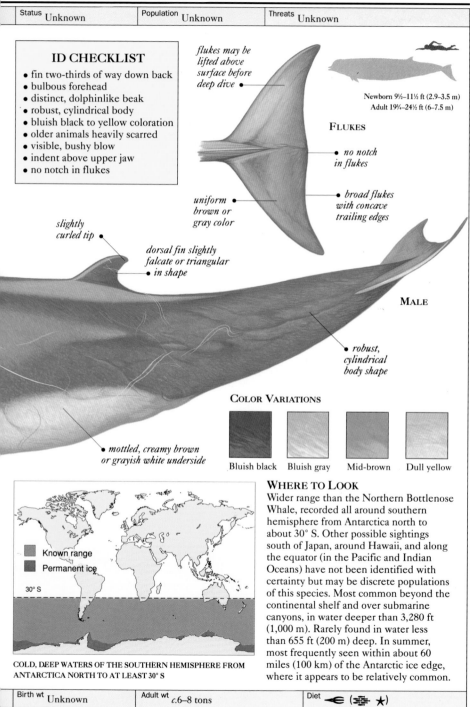

ID CHECKLIST

- fin two-thirds of way down back
- bulbous forehead
- distinct, dolphinlike beak
- robust, cylindrical body
- bluish black to yellow coloration
- older animals heavily scarred
- visible, bushy blow
- indent above upper jaw
- no notch in flukes

flukes may be lifted above surface before deep dive

Newborn 9½–11½ ft (2.9–3.5 m)
Adult 19¾–24½ ft (6–7.5 m)

FLUKES

no notch in flukes

broad flukes with concave trailing edges

uniform brown or gray color

slightly curled tip

dorsal fin slightly falcate or triangular in shape

MALE

robust, cylindrical body shape

mottled, creamy brown or grayish white underside

COLOR VARIATIONS

Bluish black   Bluish gray   Mid-brown   Dull yellow

Known range
Permanent ice

30° S

COLD, DEEP WATERS OF THE SOUTHERN HEMISPHERE FROM ANTARCTICA NORTH TO AT LEAST 30° S

WHERE TO LOOK

Wider range than the Northern Bottlenose Whale, recorded all around southern hemisphere from Antarctica north to about 30° S. Other possible sightings south of Japan, around Hawaii, and along the equator (in the Pacific and Indian Oceans) have not been identified with certainty but may be discrete populations of this species. Most common beyond the continental shelf and over submarine canyons, in water deeper than 3,280 ft (1,000 m). Rarely found in water less than 655 ft (200 m) deep. In summer, most frequently seen within about 60 miles (100 km) of the Antarctic ice edge, where it appears to be relatively common.

Birth wt Unknown	Adult wt c.6–8 tons	Diet

Family ZIPHIIDAE	Species *Mesoplodon* sp. 'A'	Habitat 〜〜

UNIDENTIFIED BEAKED WHALE

This account is based on very sketchy data, or on supposition from characteristics common to the better-known *Mesoplodon* species, so it should be taken as extremely tentative. Unlike most other beaked whales, which are known mainly from strandings, this possible new species is known only from about 30 positive sightings at sea. Unfortunately, until stranded or dead specimens are available for examination, it cannot be named. It appears to have 2 distinct color forms: one (presumably the adult male) has a very conspicuous color pattern, with a broad white or cream-colored swathe that contrasts sharply with the rest of the dark body; the other (presumably the female and juvenile) is uniformly gray-brown. The male is believed to be the larger of the two sexes and its markings should make it fairly easy to identify at sea. It may be impossible to identify a solitary female; so far, all positive sightings of females have been based on the presence of at least one male. No erupted teeth have been seen, but there is likely to be a single pair in adult males and none in females or juveniles.

• **OTHER NAMES** None.

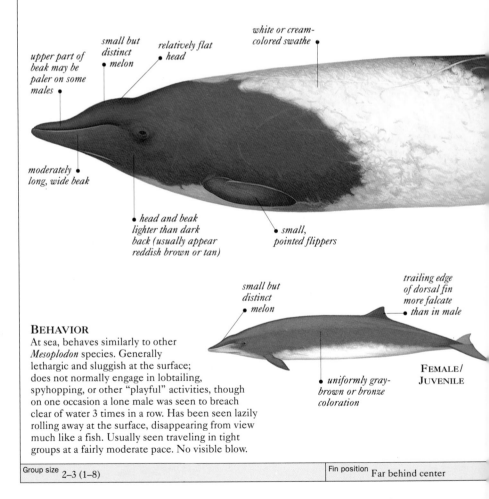

upper part of beak may be paler on some males •

small but distinct • melon

relatively flat • head

white or cream-colored swathe •

moderately • long, wide beak

• head and beak lighter than dark back (usually appear reddish brown or tan)

• small, pointed flippers

small but distinct • melon

trailing edge of dorsal fin more falcate • than in male

FEMALE/JUVENILE

• uniformly gray-brown or bronze coloration

BEHAVIOR

At sea, behaves similarly to other *Mesoplodon* species. Generally lethargic and sluggish at the surface; does not normally engage in lobtailing, spyhopping, or other "playful" activities, though on one occasion a lone male was seen to breach clear of water 3 times in a row. Has been seen lazily rolling away at the surface, disappearing from view much like a fish. Usually seen traveling in tight groups at a fairly moderate pace. No visible blow.

Group size 2–3 (1–8)	Fin position Far behind center

Status Unknown	Population Unknown	Threats Unknown

ID CHECKLIST

- dark and light coloration
- white-cream-colored swathe
- moderately long beak
- relatively flat head
- small but distinct melon
- sexes differ considerably
- small, tight groups
- lethargic and sluggish
- little aerial behavior

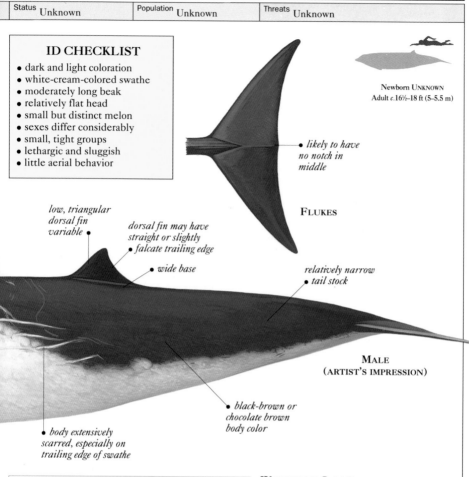

Newborn UNKNOWN
Adult *c*.16½–18 ft (5–5.5 m)

• *likely to have no notch in middle*

FLUKES

low, triangular dorsal fin variable •

dorsal fin may have straight or slightly • falcate trailing edge

• *wide base*

relatively narrow • tail stock

MALE
(ARTIST'S IMPRESSION)

• *black-brown or chocolate brown body color*

• *body extensively scarred, especially on trailing edge of swathe*

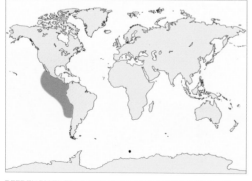

DEEP WARM WATERS OF THE EASTERN TROPICAL PACIFIC, USUALLY OFFSHORE

WHERE TO LOOK

Map shows area within which most observations have been made but, with so few records, it does not necessarily give a true representation of the real distribution of the species. One of the most frequently sighted *Mesoplodon* species in the offshore eastern tropical Pacific. So far reported off the coasts of Guatemala, El Salvador, Mexico, Costa Rica, Nicaragua, Panama, Colombia, Ecuador, and Peru. Most sightings from deep offshore waters, and all have been in areas of very warm water, around 80° F (27° C). Although its status is unknown, the number of sightings suggests that it is not particularly rare for a beaked whale.

Birth wt Unknown	Adult wt Unknown	Diet

Family ZIPHIIDAE	Species *Mesoplodon bidens*	Habitat 〰〰

SOWERBY'S BEAKED WHALE

Sowerby's Beaked Whale was the first of the beaked whales to be discovered. A lone animal was found stranded in the Moray Firth, Scotland, in 1800, and 4 years later the species was described by the English watercolor artist James Sowerby. Although it is one of the most commonly stranded *Mesoplodon* species, there have been few sightings at sea, and it is poorly known. It has one of the most northerly distributions of all the beaked whales, which should help with identification. However, parts of its range overlap with other *Mesoplodon*

species, especially Gervais' Beaked Whale (p.122), Blainville's Beaked Whale (p.120), and True's Beaked Whale (p.132), and it is likely to be difficult to distinguish from these with any certainty at sea. The position of the teeth in the male is distinctive, lying midway between the tip of the beak and corner of the mouth, although they are probably visible only at close range; the animal is also more stream-lined than most other members of the family. Females are probably unidentifiable at sea.
• **OTHER NAME** North Sea Beaked Whale.

prominent bulge located in front of blowhole (variable)

may be sandy coloration on head and beak

slate gray or bluish gray upper side

body shows limited scarring

teeth visible when mouth closed

fairly long beak

TEETH $\frac{0}{2}$

FRONT SIDE

teeth about 12 in (30 cm) from tip of lower jaw

curved trailing edges

relatively long flippers for a Mesoplodon

indentation at blowhole

may have more distinct bulge

MALE JAWBONE

beak may be more slender and dolphinlike

BEHAVIOR

Little is known. Some reports suggest head is brought out of the water at a steep angle for most surfacings. Small, bushy blow sometimes seen. Spends about 1 minute at the surface, with 4 to 6 quick breaths, followed by a long dive of 10 to 15 minutes; resurfaces up to 2,625 ft (800 m) away. Probably unobtrusive and does not approach boats. Sound of stranded animal likened to cow's mooing.

beak length variable

HEAD

Group size 1–2 (based on very little information)	Fin position Far behind center

Status Unknown	Population Unknown	Threats Unknown

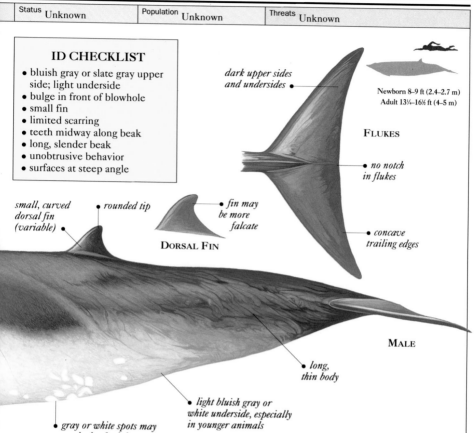

ID CHECKLIST

- bluish gray or slate gray upper side; light underside
- bulge in front of blowhole
- small fin
- limited scarring
- teeth midway along beak
- long, slender beak
- unobtrusive behavior
- surfaces at steep angle

dark upper sides and undersides •

Newborn 8–9 ft (2.4–2.7 m)
Adult 13¼–16½ ft (4–5 m)

FLUKES

• *no notch in flukes*

• *concave trailing edges*

small, curved dorsal fin (variable) • *rounded tip* • *fin may be more falcate*

DORSAL FIN

MALE

• *long, thin body*

• *light bluish gray or white underside, especially in younger animals*

• *gray or white spots may cover body of adults and sometimes younger animals*

Known range
Sightings/strandings

TEMPERATE AND SUBARCTIC WATERS IN THE EASTERN AND WESTERN NORTH ATLANTIC

WHERE TO LOOK

Known mainly from about 100 strandings. Most records from eastern North Atlantic, especially around Britain. Area west of Norway is probably center of distribution. May occur in the Mediterranean, as there is a report from Italy. Unlikely to live in the Baltic, as water too shallow. In western North Atlantic, known mainly from Newfoundland, Canada, and Massachusetts, but also from northern Labrador, Canada, and a single record from Florida. Little known about migrations; most northerly animals may migrate with advancing and retreating ice, and some populations may move toward coasts during summer. Year-round strandings, especially July to September. Probably lives some distance offshore.

Birth wt *c.*375 lb (170 kg)	Adult wt 1–1.3 tons	Diet

Family ZIPHIIDAE	Species *Mesoplodon bowdoini*	Habitat 〰〰

ANDREWS' BEAKED WHALE

Andrews' Beaked Whale is known from just over 20 strandings. It is probably extremely difficult to identify at sea, and even stranded animals have been misidentified in the past. It is considered by many experts to be the southern version of Hubbs' Beaked Whale (p.118); indeed, there are sufficient cranial and pigmentation similarities to suggest that Hubbs' Beaked Whale may be a subspecies of Andrews' Beaked Whale. The teeth are an interesting feature of this species: they are wide and flat, located on the top of a highly arched mouth line, and protrude outside the mouth in adult males; the teeth do not erupt in females and juveniles. Confusion in the Australasian region is most likely with other *Mesoplodon* species, such as Blainville's (p.120), Ginkgo-toothed (p.124), Gray's (p.126), Hector's (p.128), and Strap-toothed (p.130) Beaked Whales.

• **OTHER NAMES** Splay-toothed Beaked Whale, Bowdoin's Beaked Whale, Deepcrest Beaked Whale.

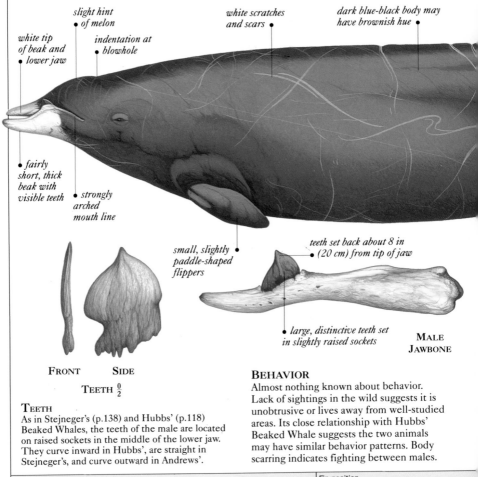

slight hint
• of melon

white tip
of beak and
• lower jaw

indentation at
• blowhole

white scratches
and scars •

dark blue-black body may
have brownish hue •

• fairly
short, thick
beak with
visible teeth

• strongly
arched
mouth line

small, slightly •
paddle-shaped
flippers

teeth set back about 8 in
• (20 cm) from tip of jaw

• large, distinctive teeth set
in slightly raised sockets

**MALE
JAWBONE**

FRONT   SIDE

TEETH $\frac{0}{2}$

TEETH
As in Stejneger's (p.138) and Hubbs' (p.118) Beaked Whales, the teeth of the male are located on raised sockets in the middle of the lower jaw. They curve inward in Hubbs', are straight in Stejneger's, and curve outward in Andrews'.

BEHAVIOR
Almost nothing known about behavior. Lack of sightings in the wild suggests it is unobtrusive or lives away from well-studied areas. Its close relationship with Hubbs' Beaked Whale suggests the two animals may have similar behavior patterns. Body scarring indicates fighting between males.

Group size Unknown	Fin position Far behind center

Status Unknown	Population Unknown	Threats Unknown

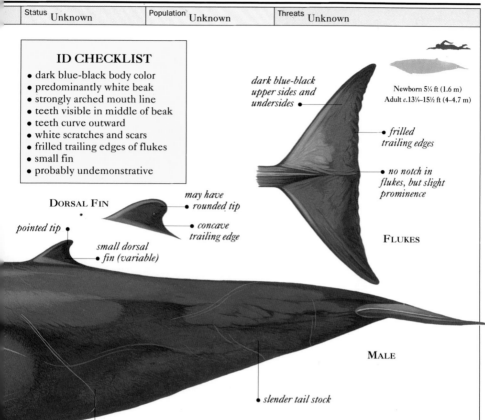

ID CHECKLIST
- dark blue-black body color
- predominantly white beak
- strongly arched mouth line
- teeth visible in middle of beak
- teeth curve outward
- white scratches and scars
- frilled trailing edges of flukes
- small fin
- probably undemonstrative

dark blue-black upper sides and undersides •

Newborn 5¼ ft (1.6 m)
Adult *c.*13¾–15½ ft (4–4.7 m)

• *frilled trailing edges*

• *no notch in flukes, but slight prominence*

FLUKES

DORSAL FIN

may have • *rounded tip*

pointed tip •

• *concave trailing edge*

small dorsal • *fin (variable)*

MALE

• *slender tail stock*

• *spindle-shaped body*

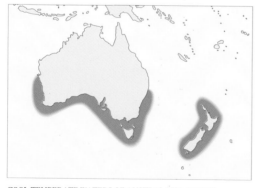

COOL TEMPERATE WATERS OF AUSTRALASIA, IN NEW ZEALAND AND ALONG THE SOUTHERN COAST OF AUSTRALIA

WHERE TO LOOK
There are too few records to be certain about distribution. Known only from strandings along the southern coast of Australia, including Tasmania, and New Zealand. Identification of a single specimen found on Kerguelen Island, in the extreme southern Indian Ocean, in 1973 is doubted by some experts, so extension of range outside Australasian region is uncertain. Several other records originally published as belonging to this species later proved to be mis-identifications; conversely, genuine specimens of Andrews' Beaked Whale may not have been recognized as such because of identification problems. Current picture of distribution may also be due to more efficient location and recording of stranded animals in New Zealand and Australia than elsewhere.

Birth wt Unknown	Adult wt 1–1.5 tons	Diet

Family ZIPHIIDAE	Species *Mesoplodon carlhubbsi*	Habitat 〰〰

HUBBS' BEAKED WHALE

The male Hubbs' Beaked Whale is one of the few beaked whales that can be positively identified at sea, although there has been only a single probable sighting (near La Jolla, California). The raised white "cap," stocky white beak, strongly arched lower jaw, and 2 massive teeth (which are clearly visible when the mouth is closed) are all distinctive. A tangle of white scars, sometimes up to 6½ ft (2 m) long, is also characteristic. At a distance, it may be confused with the Minke Whale (p.56), which has a similarly shaped and positioned dorsal fin, or the male Cuvier's Beaked Whale (p.142),

which also has a white head. Blainville's (p.120), Ginkgo-toothed (p.124), and Stejneger's (p.138) Beaked Whales are also fairly similar and overlap in range, but lack the white "cap" of Hubbs'. Females and juveniles are probably impossible to identify at sea; they have medium gray upper sides, lighter gray sides, and white undersides; their teeth do not erupt. Some experts believe that Hubbs' may be a subspecies of Andrews' Beaked Whale (p.116), because of various cranial and pigmentation similarities.
• **OTHER NAME** Arch-beaked Whale.

small light spots
• over much of body

strongly arched
• mouth line

white "cap"
• around blowhole

massive teeth
• visible

long,
stocky beak,
usually
white

relatively •
small flippers

flattened teeth set back
• from tip of jaw

head darker
than tip of
beak and
• lower jaw

FRONT   SIDE

MALE JAWBONE

TEETH $\frac{0}{2}$

mouth forms gentle,
S-shaped curve •

teeth do
not erupt •

BEHAVIOR
With only a single possible sighting, very little known about behavior. Remarkable degree of scarring suggests considerable aggression between males. Presumably, it is shy and unobtrusive like other *Mesoplodon* species. Believed to lift its head clear of the water when surfacing to breathe.

• longer, slimmer
beak than male

FEMALE

Group size Unknown	Fin position Far behind center

| Status Unknown | Population Unknown | Threats |

ID CHECKLIST

- raised white "cap"
- tangle of scars on body
- front half of beak white
- dark body color
- stocky, elongated beak
- massive teeth visible
- strongly arched lower jaw
- small, falcate fin
- probably undemonstrative

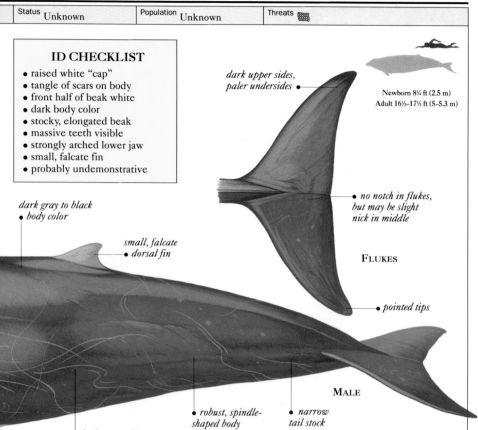

dark upper sides, paler undersides

Newborn 8¼ ft (2.5 m)
Adult 16½–17½ ft (5–5.3 m)

no notch in flukes, but may be slight nick in middle

FLUKES

pointed tips

dark gray to black body color

small, falcate dorsal fin

MALE

body covered in scratches and scars

robust, spindle-shaped body

narrow tail stock

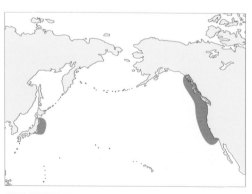

COLD TEMPERATE WATERS OF THE EASTERN AND WESTERN NORTH PACIFIC

WHERE TO LOOK

In the eastern North Pacific, found roughly between 33° N (part of a skull was brought up by a submersible southwest of San Clemente Island, California) and 54° N (Prince Rupert, British Columbia, Canada); distribution may be related to confluence of the subarctic and Californian current systems. Most records from California. More restricted range in the western North Pacific, with a few records from around the fishing town of Ayukawa, Honshu, Japan, where the warm, north-flowing Kuroshio Current meets the cold, south-flowing Oyashio Current in the southern Sea of Japan. Unlikely to be common in Japanese waters. Map shows stranding areas, but probably pelagic and may stretch right across the North Pacific.

| Birth wt Unknown | Adult wt 1–1.5 tons | Diet |

Family ZIPHIIDAE	Species *Mesoplodon densirostris*	Habitat 〰

BLAINVILLE'S BEAKED WHALE

The male Blainville's Beaked Whale is one of the oddest-looking of all cetaceans. It has a pair of massive teeth that grow from substantial bulges in its lower jaw, like a couple of horns; these may be so encrusted with barnacles that the animal appears to have 2 dark-colored pompons on top of its head. This feature makes it relatively easy to identify at sea, although it is generally inconspicuous and difficult to find; it is known mainly from strandings. The flattened forehead and large spots all over its body, possibly made by the teeth of Cookie-cutter Sharks and parasites, are also characteristic.

The teeth do not erupt in females and, although they have similar (albeit less distinctive) bulges in the lower jaw, they are probably extremely difficult to distinguish from other *Mesoplodon* females. The species was named *densirostris*, meaning "dense beak," because the original description was based on a small but extremely heavy piece of upper jaw; it was later determined that Blainville's Beaked Whale has the densest bones in the animal kingdom.

• **OTHER NAMES** Atlantic Beaked Whale, Dense Beaked Whale, Tropical Beaked Whale.

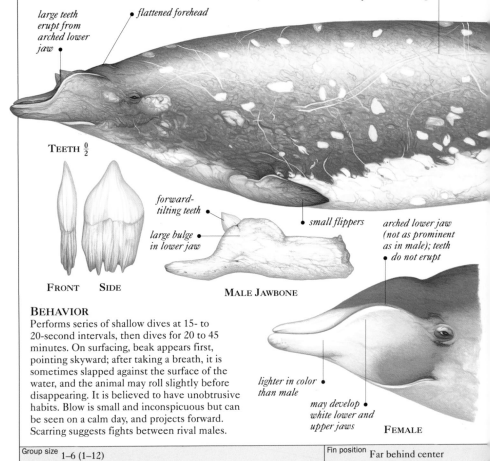

dark bluish gray upper side and sides may darken with age •

large teeth erupt from arched lower jaw •

• *flattened forehead*

TEETH $\frac{0}{2}$

forward-tilting teeth •

large bulge in lower jaw •

• *small flippers*

arched lower jaw (not as prominent as in male); teeth • *do not erupt*

FRONT SIDE

MALE JAWBONE

BEHAVIOR
Performs series of shallow dives at 15- to 20-second intervals, then dives for 20 to 45 minutes. On surfacing, beak appears first, pointing skyward; after taking a breath, it is sometimes slapped against the surface of the water, and the animal may roll slightly before disappearing. It is believed to have unobtrusive habits. Blow is small and inconspicuous but can be seen on a calm day, and projects forward. Scarring suggests fights between rival males.

lighter in color than male •

may develop • *white lower and upper jaws*

FEMALE

Group size 1–6 (1–12)	Fin position Far behind center

Status Unknown	Population Unknown	Threats Unknown

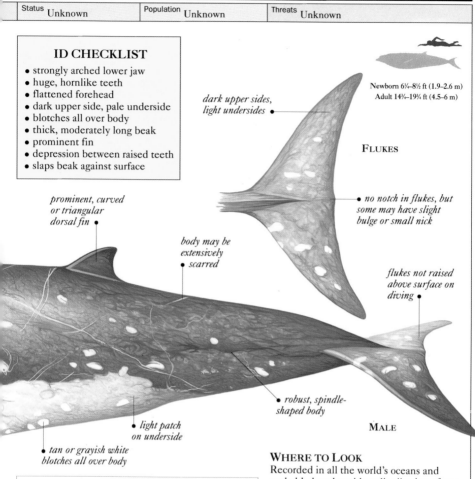

ID CHECKLIST
- strongly arched lower jaw
- huge, hornlike teeth
- flattened forehead
- dark upper side, pale underside
- blotches all over body
- thick, moderately long beak
- prominent fin
- depression between raised teeth
- slaps beak against surface

dark upper sides, light undersides

Newborn 6¼–8½ ft (1.9–2.6 m)
Adult 14¾–19¾ ft (4.5–6 m)

FLUKES

prominent, curved or triangular dorsal fin

body may be extensively scarred

no notch in flukes, but some may have slight bulge or small nick

flukes not raised above surface on diving

robust, spindle-shaped body

MALE

light patch on underside

tan or grayish white blotches all over body

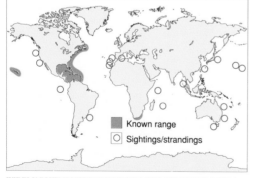

■ Known range
▢ Sightings/strandings

WIDELY DISTRIBUTED IN WARM TEMPERATE AND TROPICAL WATERS, PRIMARILY ON THE ATLANTIC COAST OF THE USA

WHERE TO LOOK
Recorded in all the world's oceans and probably has the widest distribution of any *Mesoplodon* species. Atlantic coast of USA seems to be main area of concentration and, to a much lesser extent, South Africa; small schools have been sighted in Hawaii, especially off the Waianae coast of Oahu. Widely distributed records from other parts of the world, but these are small in number. A female was recently stranded in Britain. Seems to prefer deep waters and is thought to be one of the most pelagic of the beaked whales, because it strands on oceanic islands as often as on mainland. Seems to avoid polar waters. May be one of the most common members of the family, but sightings are rare, probably because of its distance from land.

Birth wt c.130 lb (60 kg)	Adult wt c.1 ton	Diet

Family ZIPHIIDAE	Species *Mesoplodon europaeus*	Habitat 〰〰

GERVAIS' BEAKED WHALE

Like most of the beaked whales, Gervais' Beaked Whale is poorly known. The female is probably impossible to identify at sea, and males are likely to be exceedingly difficult. A stranded male can be identified by the single pair of teeth, which are located one-third of the way from the tip of the beak to the corner of the mouth; these are visible when the animal's mouth is closed and fit neatly into grooves in the upper "lip." The teeth do not erupt in females. Confusion is most likely to

occur with True's Beaked Whale (p.132), Cuvier's Beaked Whale (p.142), Blainville's Beaked Whale (p.120), and Sowerby's Beaked Whale (p.114). The exact position of the teeth and the prominent but narrow beak of the male Gervais' Beaked Whale are probably the best clues for identification at sea.

• OTHER NAMES Gulf Stream Beaked Whale, European Beaked Whale, Antillean Beaked Whale.

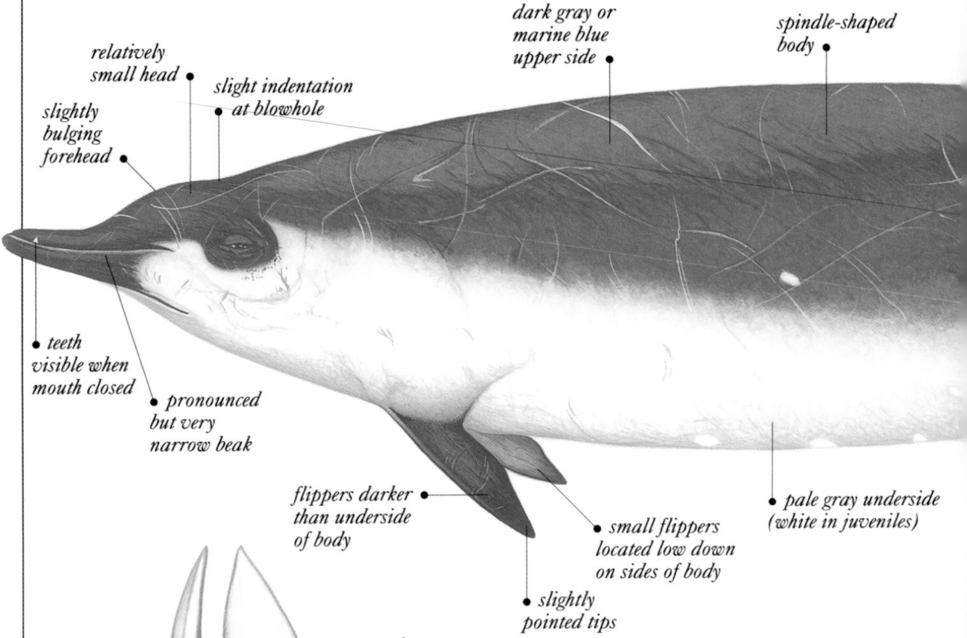

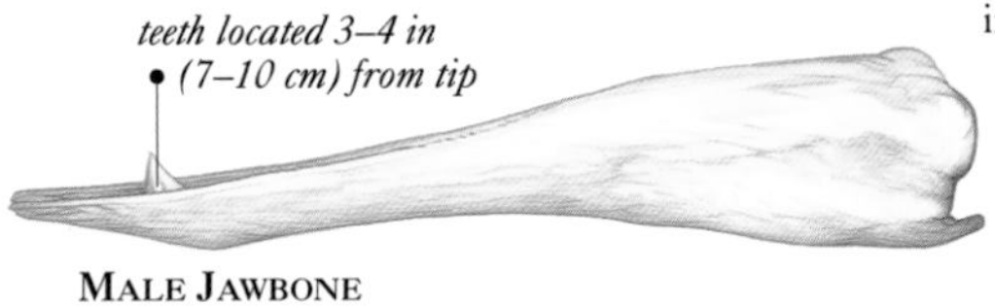

relatively small head

slight indentation at blowhole

slightly bulging forehead

dark gray or marine blue upper side

spindle-shaped body

teeth visible when mouth closed

pronounced but very narrow beak

flippers darker than underside of body

small flippers located low down on sides of body

slightly pointed tips

pale gray underside (white in juveniles)

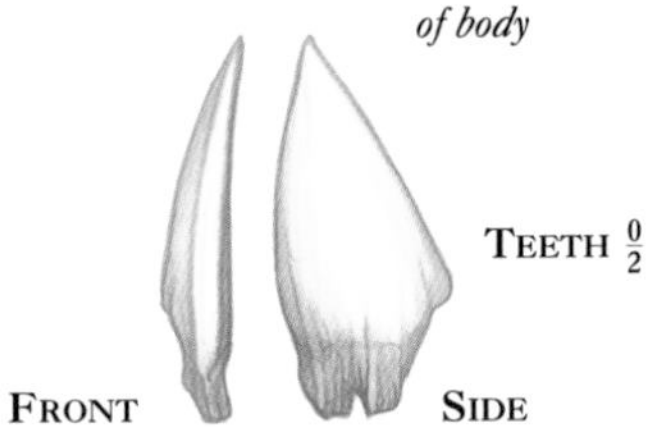

TEETH $\frac{0}{2}$

FRONT   SIDE

teeth located 3–4 in (7–10 cm) from tip

MALE JAWBONE

BEHAVIOR

Behavior in the wild is a matter for conjecture. Lack of sightings in relatively well-studied areas within range suggests it is likely to be inconspicuous. Probably a deep diver living in small groups or pairs. Some reports indicate beak breaks water first when whale surfaces to breathe. Scarring suggests fighting between males. Has been known to become entangled in fishing nets.

Group size 2–5	Fin position Far behind center

Status Unknown	Population Unknown	Threats

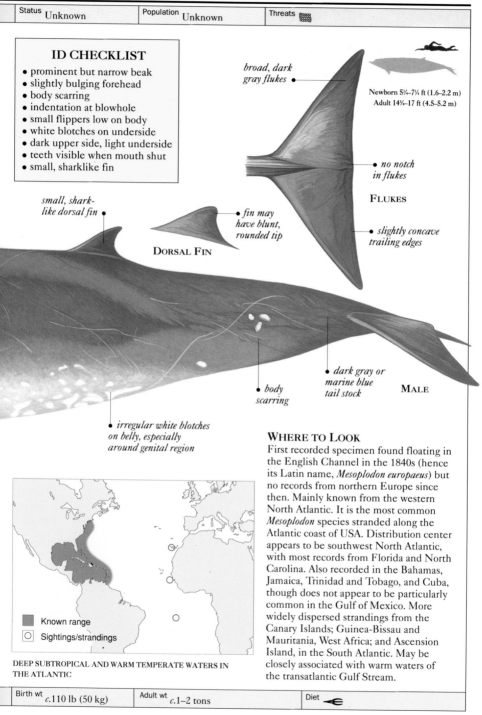

ID CHECKLIST

- prominent but narrow beak
- slightly bulging forehead
- body scarring
- indentation at blowhole
- small flippers low on body
- white blotches on underside
- dark upper side, light underside
- teeth visible when mouth shut
- small, sharklike fin

broad, dark gray flukes

Newborn 5¼–7¼ ft (1.6–2.2 m)
Adult 14¾–17 ft (4.5–5.2 m)

no notch in flukes

FLUKES

slightly concave trailing edges

small, shark-like dorsal fin

fin may have blunt, rounded tip

DORSAL FIN

dark gray or marine blue tail stock **MALE**

body scarring

irregular white blotches on belly, especially around genital region

Known range
Sightings/strandings

DEEP SUBTROPICAL AND WARM TEMPERATE WATERS IN THE ATLANTIC

WHERE TO LOOK

First recorded specimen found floating in the English Channel in the 1840s (hence its Latin name, *Mesoplodon europaeus*) but no records from northern Europe since then. Mainly known from the western North Atlantic. It is the most common *Mesoplodon* species stranded along the Atlantic coast of USA. Distribution center appears to be southwest North Atlantic, with most records from Florida and North Carolina. Also recorded in the Bahamas, Jamaica, Trinidad and Tobago, and Cuba, though does not appear to be particularly common in the Gulf of Mexico. More widely dispersed strandings from the Canary Islands; Guinea-Bissau and Mauritania, West Africa; and Ascension Island, in the South Atlantic. May be closely associated with warm waters of the transatlantic Gulf Stream.

Birth wt *c.*110 lb (50 kg)	Adult wt *c.*1–2 tons	Diet

Family ZIPHIIDAE	Species *Mesoplodon ginkgodens*	Habitat 〰〰

GINKGO-TOOTHED BEAKED WHALE

The Ginkgo-toothed Beaked Whale is very poorly known. It is named after the male's distinctive teeth, which are shaped like the leaves of the Ginkgo tree – a common tree in Japan, where the first specimens of this whale were found and described. These teeth are about 4 in (10 cm) wide, making them the widest of any known *Mesoplodon* species. Males have a uniformly dark body color that becomes darker soon after death; some experts have suggested that the color of the live animal is a marine blue. It is less heavily scarred than most other *Mesoplodon* species. However, the male has characteristic white spots and blotches around the navel; these are typically 1¼–1½ in (3–4 cm) wide and may be parasitic scars rather than true pigmentation. Females are thought to have medium gray upper sides and light gray undersides, and possibly lighter colored heads. It is probably extremely difficult to identify males at sea, and positive identification of females is likely to be impossible.

• **OTHER NAMES** Japanese Beaked Whale, Ginkgo Beaked Whale.

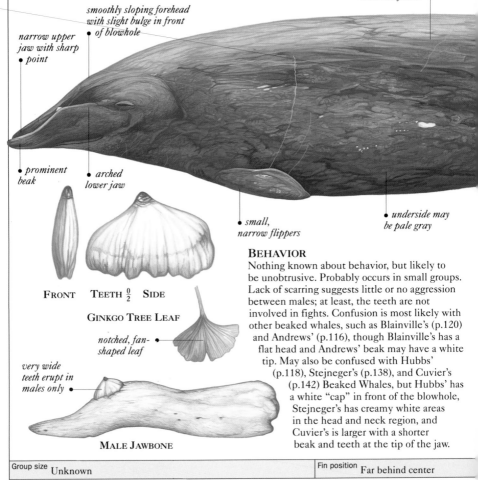

dark body color

smoothly sloping forehead with slight bulge in front of blowhole

narrow upper jaw with sharp point

prominent beak

arched lower jaw

small, narrow flippers

underside may be pale gray

FRONT TEETH ⁰⁄₂ SIDE

GINKGO TREE LEAF

notched, fan-shaped leaf

very wide teeth erupt in males only

MALE JAWBONE

BEHAVIOR

Nothing known about behavior, but likely to be unobtrusive. Probably occurs in small groups. Lack of scarring suggests little or no aggression between males; at least, the teeth are not involved in fights. Confusion is most likely with other beaked whales, such as Blainville's (p.120) and Andrews' (p.116), though Blainville's has a flat head and Andrews' beak may have a white tip. May also be confused with Hubbs' (p.118), Stejneger's (p.138), and Cuvier's (p.142) Beaked Whales, but Hubbs' has a white "cap" in front of the blowhole, Stejneger's has creamy white areas in the head and neck region, and Cuvier's is larger with a shorter beak and teeth at the tip of the jaw.

Group size Unknown	Fin position Far behind center

Status Rare	Population Unknown	Threats Unknown

ID CHECKLIST

- moderately long beak
- male body bluish black
- female body medium gray
- little or no scarring
- smoothly sloping forehead
- arched lower jaw
- flaps of skin largely cover teeth
- small, pointed dorsal fin
- teeth close to middle of jaw

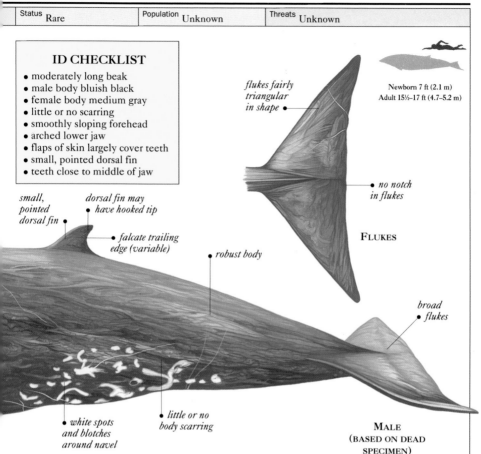

flukes fairly triangular in shape

Newborn 7 ft (2.1 m)
Adult 15½–17 ft (4.7–5.2 m)

small, pointed dorsal fin

dorsal fin may have hooked tip

falcate trailing edge (variable)

robust body

no notch in flukes

FLUKES

broad flukes

white spots and blotches around navel

little or no body scarring

MALE
(BASED ON DEAD SPECIMEN)

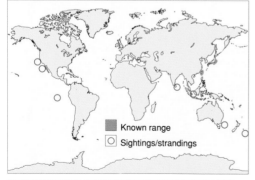

WARM TEMPERATE AND TROPICAL WATERS IN THE PACIFIC AND INDIAN OCEANS

Known range

Sightings/strandings

WHERE TO LOOK

Known from only a very small number of widely distributed strandings. Primarily recorded in the North Pacific and may be most common in the western North Pacific, especially off the coasts of Japan. Apparently also occurs in the South Pacific and Indian Oceans. Seems to prefer warm temperate to tropical regions, and normal habitat is assumed to be in deep water. Appears to be uncommon; however, it may simply live away from major shipping lanes and outside well-studied areas, and may live so far from land that few specimens survive long enough after death to be washed ashore.

Birth wt Unknown	Adult wt *c.*1.5–2 tons	Diet

Family ZIPHIIDAE	Species *Mesoplodon grayi*	Habitat 〰〰

GRAY'S BEAKED WHALE

The male Gray's Beaked Whale is difficult, but not impossible, to identify at sea; the straight mouth line and white slender beak are both distinctive and are frequently visible above the surface. There have been a number of confirmed sightings, mainly from the southern Indian Ocean, although most information that is available is from stranded animals. The male has 2 fairly small, triangular teeth, which are set back from the tip of the beak and are visible when the mouth is closed; these do not usually erupt in the female. Both sexes have rows of tiny vestigial teeth in the upper jaw; these are embedded in the gum, rather than the bone, behind the main teeth, and are usually visible. Many beaked whales have vestigial teeth, but it is unusual for them to erupt. Their presence prompted a proposal to place Gray's Beaked Whale in a new genus, *Oulodon*, but this has never been widely accepted. From the little evidence available, this species may be social, which is unusual for beaked whales. Females and juveniles are probably impossible to identify at sea.

• **OTHER NAMES** Scamperdown Whale, Southern Beaked Whale.

small head with flat forehead •

indentation at • blowhole

white beak, front of forehead, and throat (variable) •

dark bluish gray, brownish • gray, or black upper side

• long, slender beak

fairly • straight mouth line

ΦΦΦΦΦΦΦΦΦΦΦΦΦΦΦΦ

VESTIGIAL TEETH

short, wide • flippers

• white or yellowish spots on underside and sides

FRONT   **SIDE**

TEETH $\frac{34-44}{2}$

BEHAVIOR

Limited number of sightings suggests it may be more conspicuous at the surface than other beaked whales: seems to be more active and may live in larger groups. Observed singly, in pairs, and in small groups, but a mass stranding of 28 animals in the Chatham Islands, east of New Zealand, in 1874 suggests that fairly large numbers may be encountered together. Has been seen breaching at a shallow angle, lifting entire body as far as flukes out of the water. When swimming at speed, also observed porpoising through the water, making low, arc-shaped leaps, rather like rightwhale dolphins (see p.168). Typically pokes white beak out of water as it surfaces to blow.

serrated • edge

• fairly small teeth 8–9½ in (20–24 cm) behind tip of beak

MALE JAWBONE

Group size 2–6 (1–10), 28 stranded together on a single occasion	Fin position Far behind center

Status Unknown	Population Unknown	Threats Unknown

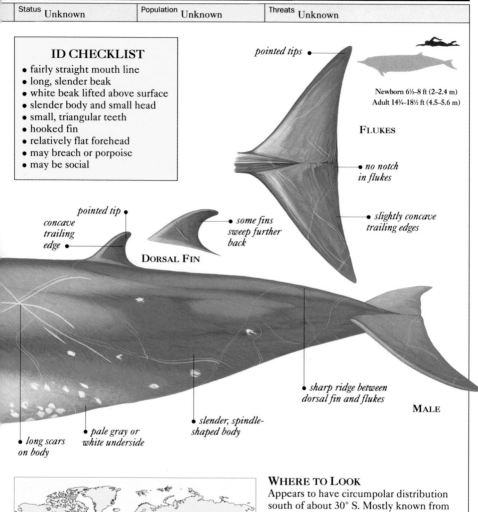

ID CHECKLIST

- fairly straight mouth line
- long, slender beak
- white beak lifted above surface
- slender body and small head
- small, triangular teeth
- hooked fin
- relatively flat forehead
- may breach or porpoise
- may be social

pointed tips •

Newborn 6½–8 ft (2–2.4 m)
Adult 14¾–18½ ft (4.5–5.6 m)

FLUKES

• *no notch in flukes*

• *slightly concave trailing edges*

pointed tip •
concave trailing edge •

• *some fins sweep further back*

DORSAL FIN

• *sharp ridge between dorsal fin and flukes*

MALE

• *slender, spindle-shaped body*

• *pale gray or white underside*

• *long scars on body*

WHERE TO LOOK

Appears to have circumpolar distribution south of about 30° S. Mostly known from strandings in New Zealand, though increasing number of records from Tierra del Fuego (in southern South America), the Falkland Islands, South Africa, Australia, and the Chatham Islands. Only 1 record in the northern hemisphere: an animal that stranded on the North Sea coast of the Netherlands in 1927; since there have been no other records in the region before or since, this was probably a stray. Significant number of sightings from a deep-water area south of Madagascar. Possible sighting in the Seychelles in early 1980s of 3 adults and a juvenile.

Known range
Possible range
○ Sightings/strandings

30° S

COOL TEMPERATE WATERS OF THE SOUTHERN HEMISPHERE, SOUTH OF 30° S

Birth wt Unknown	Adult wt 1–1.5 tons	Diet

Family ZIPHIIDAE	Species *Mesoplodon hectori*	Habitat Unknown

HECTOR'S BEAKED WHALE

This is one of the smaller members of the family, with the smallest skull of any *Mesoplodon* species. It was first discovered in 1866, but until 1975 only 7 decomposed specimens were known; all were from the southern hemisphere. The first recognizable adult male was not found until 1978. Nowadays, it is known from more than 20 specimens, 4 of which were stranded in California, suggesting a more extensive range into the North Pacific. Few have been examined in the flesh, since most are known only from skeletons or skulls. However, there have been 2 probable sightings, both of which

were in California: in July 1976, a pair of animals was photographed off Catalina Island, and in September 1978, a second pair was spotted just over 50 miles (80 km) west of San Diego. This is an extremely difficult species to recognize at sea, although the combination of its small size and the shape and position of its teeth (which erupt only in males) may be distinctive.
• **OTHER NAMES** New Zealand Beaked Whale, Skew-beaked Whale.

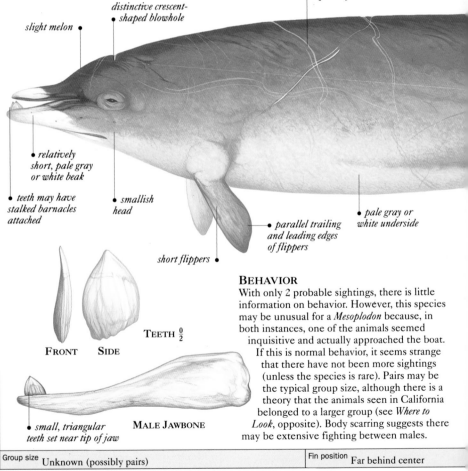

body covered in scratches and sometimes oval scars, • especially in males

distinctive crescent-• shaped blowhole

slight melon •

• relatively short, pale gray or white beak

• teeth may have stalked barnacles attached

• smallish head

• parallel trailing and leading edges of flippers

• pale gray or white underside

short flippers •

TEETH $\frac{0}{2}$

FRONT   **SIDE**

• small, triangular teeth set near tip of jaw

MALE JAWBONE

BEHAVIOR

With only 2 probable sightings, there is little information on behavior. However, this species may be unusual for a *Mesoplodon* because, in both instances, one of the animals seemed inquisitive and actually approached the boat. If this is normal behavior, it seems strange that there have not been more sightings (unless the species is rare). Pairs may be the typical group size, although there is a theory that the animals seen in California belonged to a larger group (see *Where to Look*, opposite). Body scarring suggests there may be extensive fighting between males.

Group size Unknown (possibly pairs)	Fin position Far behind center

Status Unknown	Population Unknown	Threats Unknown

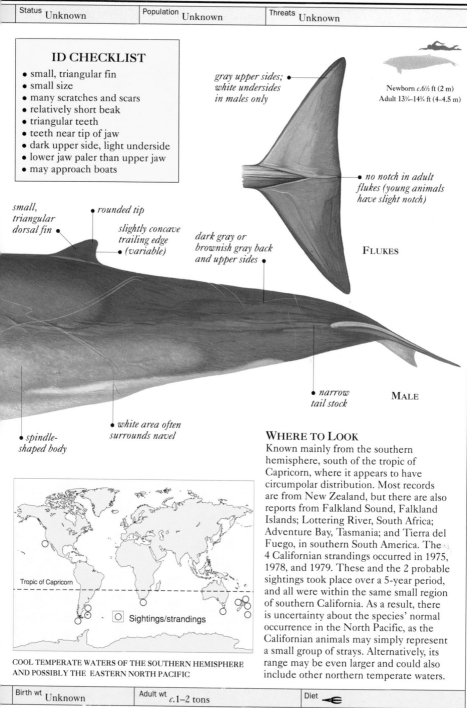

ID CHECKLIST

- small, triangular fin
- small size
- many scratches and scars
- relatively short beak
- triangular teeth
- teeth near tip of jaw
- dark upper side, light underside
- lower jaw paler than upper jaw
- may approach boats

gray upper sides; • white undersides in males only

Newborn *c*.6½ ft (2 m)
Adult 13¼–14¾ ft (4–4.5 m)

• no notch in adult flukes (young animals have slight notch)

FLUKES

small, triangular dorsal fin •

• rounded tip

slightly concave trailing edge (*variable*) •

dark gray or brownish gray back and upper sides •

• slightly concave trailing edge

• narrow tail stock

MALE

• white area often surrounds navel

• spindle-shaped body

WHERE TO LOOK

Known mainly from the southern hemisphere, south of the tropic of Capricorn, where it appears to have circumpolar distribution. Most records are from New Zealand, but there are also reports from Falkland Sound, Falkland Islands; Lottering River, South Africa; Adventure Bay, Tasmania; and Tierra del Fuego, in southern South America. The 4 Californian strandings occurred in 1975, 1978, and 1979. These and the 2 probable sightings took place over a 5-year period, and all were within the same small region of southern California. As a result, there is uncertainty about the species' normal occurrence in the North Pacific, as the Californian animals may simply represent a small group of strays. Alternatively, its range may be even larger and could also include other northern temperate waters.

Tropic of Capricorn

☐ Sightings/strandings

COOL TEMPERATE WATERS OF THE SOUTHERN HEMISPHERE AND POSSIBLY THE EASTERN NORTH PACIFIC

Birth wt Unknown	Adult wt *c*.1–2 tons	Diet

Family ZIPHIIDAE	Species *Mesoplodon layardii*	Habitat 〰〰

STRAP-TOOTHED WHALE

One of the largest of the beaked whales, the Strap-toothed Whale is also one of the few species of the genus *Mesoplodon* that can be readily identified at sea. The adult male has 2 extraordinary teeth that grow from its lower jaw, curling upward and backward and over the top of its upper jaw. In older animals, the teeth sometimes grow to a length of 12 in (30 cm) or more, and may meet in the middle. They act like a muzzle, preventing the whale from opening its mouth properly; however, it can still feed by using its beak like a vacuum cleaner and, indeed, the teeth may act as "guardrails" to keep the food on a direct path to the throat. The teeth

do not erupt in females, making identification at sea virtually impossible (although an adult male may be present in many groups). Young males have smaller, more triangular teeth than adults.

• **OTHER NAMES** Layard's Beaked Whale, Strap-tooth Beaked Whale.

*dark areas paler
• than on adult*

*light and dark pattern same
• as on adult, but in reverse*

CALF

*sloping forehead
with slightly
bulging melon •*

*black "face
• mask"*

*long
teeth, sometimes
with seaweed
attached*

*areas of white and •
gray may be yellowish,
especially after death*

*small,
narrow
flippers •*

BEHAVIOR

Rarely seen in the wild. May bask at the surface on calm, sunny days, but generally hard to approach, especially in large vessels. Flukes do not normally show above the surface at start of dive. Limited observations suggest that it sinks slowly beneath the surface, barely creating a ripple, then rises and blows again 490–655 ft (150–200 m) away; or it dives with a characteristic sideways roll, showing a single flipper above the surface, and reappears some distance away. Typical dive time is 10 to 15 minutes. Believed to break the surface first with its beak and then head when rising to breathe. Scarring suggests fighting between males.

*both teeth curl
over upper jaw
and meet in middle*

*mouth cannot
open properly*

OLDER MALE
(HEAD-ON)

TEETH $\frac{0}{2}$ FRONT   SIDE

*backward-tilting •
teeth 12 in (30 cm) or
more from tip of jaw*

MALE JAWBONE

Group size 1–3	Fin position Far behind center

Status Unknown	Population Unknown	Threats Unknown

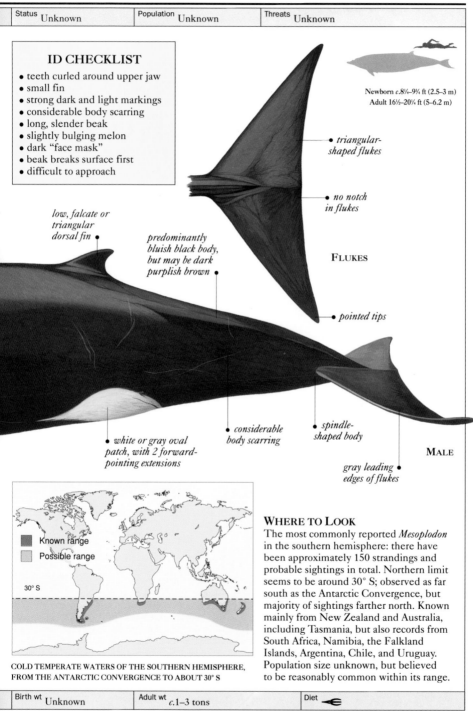

ID CHECKLIST

- teeth curled around upper jaw
- small fin
- strong dark and light markings
- considerable body scarring
- long, slender beak
- slightly bulging melon
- dark "face mask"
- beak breaks surface first
- difficult to approach

Newborn *c.*8¼–9¾ ft (2.5–3 m)
Adult 16½–20¼ ft (5–6.2 m)

- *triangular-shaped flukes*

- *no notch in flukes*

FLUKES

- *pointed tips*

low, falcate or triangular dorsal fin •

predominantly bluish black body, but may be dark purplish brown •

- *white or gray oval patch, with 2 forward-pointing extensions*

- *considerable body scarring*

- *spindle-shaped body*

gray leading edges of flukes •

MALE

Known range
Possible range

30° S

COLD TEMPERATE WATERS OF THE SOUTHERN HEMISPHERE,
FROM THE ANTARCTIC CONVERGENCE TO ABOUT 30° S

WHERE TO LOOK

The most commonly reported *Mesoplodon* in the southern hemisphere: there have been approximately 150 strandings and probable sightings in total. Northern limit seems to be around 30° S; observed as far south as the Antarctic Convergence, but majority of sightings farther north. Known mainly from New Zealand and Australia, including Tasmania, but also records from South Africa, Namibia, the Falkland Islands, Argentina, Chile, and Uruguay. Population size unknown, but believed to be reasonably common within its range.

Birth wt Unknown	Adult wt *c.*1–3 tons	Diet

Family ZIPHIIDAE	Species *Mesoplodon mirus*	Habitat 〰〰

TRUE'S BEAKED WHALE

True's Beaked Whale is little known and has not yet been positively identified at sea. Its range overlaps with those of several *Mesoplodon* species, but it may be distinguished by the male's 2 small teeth at the tip of the lower jaw. Cuvier's Beaked Whale (p.142) has teeth in a similar position, but its lighter head, less prominent beak, and larger size should be distinctive. In female True's Beaked Whales, the teeth are concealed beneath the gums; female and juvenile animals are probably unidentifiable, unless stranded specimens can be examined closely. There may be 2 forms of this species: the best-known one lives in the North Atlantic, the other is found in parts of the southern hemisphere. There are slight cranial and pigmentation differences between them. True's Beaked Whale is known from about 40 specimens, 75 percent of which are from the North Atlantic. In 1913 the American biologist Frederick True named the species *mirus*, meaning "wonderful."

• **OTHER NAME** Wonderful Beaked Whale.

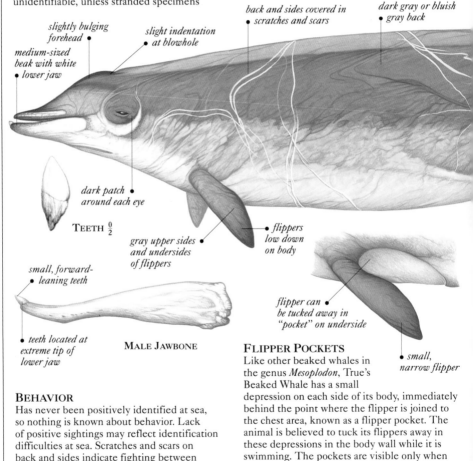

slightly bulging forehead •

slight indentation • at blowhole

medium-sized beak with white • lower jaw

back and sides covered in • scratches and scars

dark gray or bluish • gray back

dark patch • around each eye

TEETH $\frac{0}{2}$

gray upper sides and undersides of flippers •

• flippers low down on body

small, forward- • leaning teeth

• teeth located at extreme tip of lower jaw

MALE JAWBONE

flipper can • be tucked away in "pocket" on underside

• small, narrow flipper

BEHAVIOR

Has never been positively identified at sea, so nothing is known about behavior. Lack of positive sightings may reflect identification difficulties at sea. Scratches and scars on back and sides indicate fighting between males. Likely to be a deep diver.

FLIPPER POCKETS

Like other beaked whales in the genus *Mesoplodon*, True's Beaked Whale has a small depression on each side of its body, immediately behind the point where the flipper is joined to the chest area, known as a flipper pocket. The animal is believed to tuck its flippers away in these depressions in the body wall while it is swimming. The pockets are visible only when a stranded animal can be examined closely.

Group size Unknown	Fin position Far behind center

Status Unknown	Population Unknown	Threats Unknown

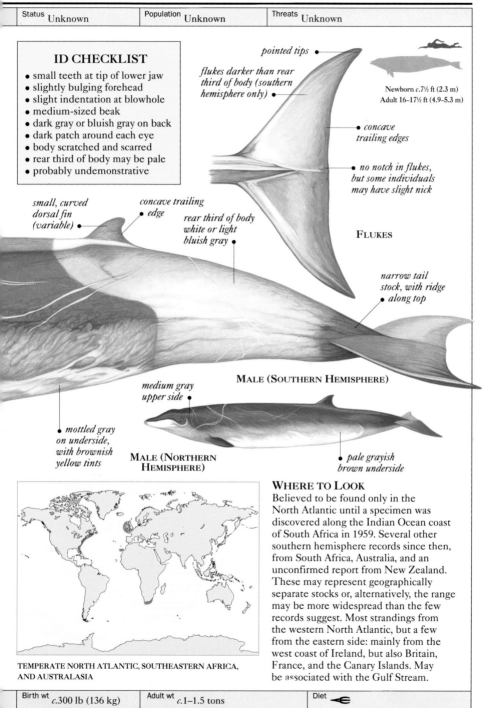

ID CHECKLIST

- small teeth at tip of lower jaw
- slightly bulging forehead
- slight indentation at blowhole
- medium-sized beak
- dark gray or bluish gray on back
- dark patch around each eye
- body scratched and scarred
- rear third of body may be pale
- probably undemonstrative

pointed tips •

flukes darker than rear third of body (southern hemisphere only) •

Newborn *c.*7½ ft (2.3 m)
Adult 16–17½ ft (4.9–5.3 m)

• *concave trailing edges*

• *no notch in flukes, but some individuals may have slight nick*

small, curved dorsal fin (variable) •

concave trailing edge •

rear third of body white or light bluish gray •

FLUKES

narrow tail stock, with ridge along top •

MALE (SOUTHERN HEMISPHERE)

medium gray upper side •

• *mottled gray on underside, with brownish yellow tints*

MALE (NORTHERN HEMISPHERE)

• *pale grayish brown underside*

WHERE TO LOOK

Believed to be found only in the North Atlantic until a specimen was discovered along the Indian Ocean coast of South Africa in 1959. Several other southern hemisphere records since then, from South Africa, Australia, and an unconfirmed report from New Zealand. These may represent geographically separate stocks or, alternatively, the range may be more widespread than the few records suggest. Most strandings from the western North Atlantic, but a few from the eastern side: mainly from the west coast of Ireland, but also Britain, France, and the Canary Islands. May be associated with the Gulf Stream.

TEMPERATE NORTH ATLANTIC, SOUTHEASTERN AFRICA, AND AUSTRALASIA

Birth wt *c.*300 lb (136 kg)	Adult wt *c.*1–1.5 tons	Diet

Family ZIPHIIDAE	Species *Mesoplodon pacificus*	Habitat Unknown

LONGMAN'S BEAKED WHALE

Longman's Beaked Whale is probably the least known of the world's whales, since research is based on only 2 weathered skulls. These are large for a *Mesoplodon*, suggesting an animal around 23 ft (7 m) long. They are sufficiently distinctive for some experts to place the whale in a genus of its own, called *Indopacetus*, but others still dispute its status as a separate species. It was suggested that it might be a form of Southern Bottlenose Whale (p.110), but this is unlikely, as there are many cranial differences. Another theory is that it may be a subspecies of True's Beaked Whale (p.132), which does have a similar skull. Longman's Beaked Whale has

never been positively identified in the flesh, although there have been several possible sightings. In 1980, 2 light gray whales were seen by experienced observers near the Seychelles, in the Indian Ocean: one was estimated at 24¾ ft (7.5 m) and the other at 15 ft (4.6 m); both had elongated beaks and broad flukes with straight trailing edges. Since there is no definitive information on Longman's Beaked Whale, this illustration is strictly an artist's impression based on True's Beaked Whale and possible sightings.

• **OTHER NAMES** Pacific Beaked Whale, Indo-Pacific Beaked Whale.

skull shape
suggests distinct
• beak

• single blowhole

• lower jaw
with 2 teeth
inclined forward
at tip

• V-shaped
throat grooves

• likely to have a
spindle-shaped body

SKULL

The skull is characterized by having a single pair of teeth at the tip of the lower jaw. No teeth have been found, but small cavities indicate that they are slightly compressed and more oval than round in cross section. The Mackay skull (see *Where to Look*, opposite) is 4 ft (1.2 m) long; the Danane skull is 3½ ft (1.1 m). The Danane skull was incomplete and had to be partially reconstructed.

distinctive
lateral swelling
• on upper jaw

• 2 small cavities
indicate presence of teeth

• long beak

• hood-shaped
cheekbones

Group size Unknown	Fin position Unknown

Status Unknown	Population Unknown	Threats Unknown

ID CHECKLIST
Based on supposition

- 23 ft (7 m) or more in length
- no notch in flukes
- small fin
- small flippers set far forward
- teeth at tip of lower jaw
- elongated beak
- probably pelagic
- unobtrusive habits

IDENTIFICATION
The 2 skulls provide
enough information to
place Longman's Beaked
Whale in the family Ziphiidae:
they are asymmetrical, with distinct
beaks, and the lower jaws extend beyond
the tips of the upper jaws. There are only 2 laterally
compressed teeth in the lower jaws, placing the species
tentatively in the genus *Mesoplodon*. Characteristics of other
Mesoplodon species indicate that it probably feeds on squid
and lives in the deep waters of the open sea; it is likely to
have a spindle-shaped body, 2 V-shaped throat grooves, no
notch in its flukes, and a small dorsal fin positioned far back
on its body. However, until a complete specimen has been
found, positive identification in the wild will be impossible.

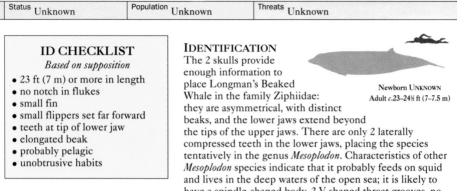

Newborn UNKNOWN
Adult *c.*23–24½ ft (7–7.5 m)

possibly small dorsal fin, positioned far back on body

unlikely to have notch in middle of flukes

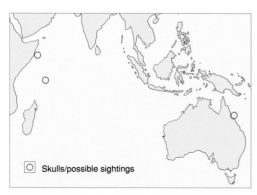

MALE
(ARTIST'S IMPRESSION)

nothing known about body color, but likely to be scratched and scarred

☐ Skulls/possible sightings

POSSIBLY DEEP TROPICAL WATERS OF THE INDIAN AND
PACIFIC OCEANS

WHERE TO LOOK
The first skull was found on a beach
near Mackay, northeastern Queensland,
Australia, in 1882; in 1926, the species was
named from this evidence by Longman.
The second skull was found in 1955 on the
floor of a fertilizer factory in Mogadishu,
Somalia; this was later traced to a beach
near Danane, on the northeast coast of
Somalia, where it had been picked up by
local fishermen. These widely separate
locations suggest an extensive range
in both the Indian and Pacific Oceans.
Several possible sightings of unidentified
beaked whales in the tropical waters of
both oceans may support this evidence.
Based on knowledge of other *Mesoplodon*
species, and the fact that it is rarely seen,
it is thought to live in deep, pelagic waters.

Birth wt Unknown	Adult wt Unknown	Diet Unknown

Family ZIPHIIDAE	Species *Mesoplodon peruvianus*	Habitat Unknown

LESSER BEAKED WHALE

The Lesser Beaked Whale is the smallest *Mesoplodon* species. It is known from only 13 specimens and a handful of possible sightings at sea. Scientists have been aware of its existence since 1976, when part of a mysterious skull was discovered at a fish market in San Andrés, Peru; the shape of the skull identified it as a *Mesoplodon*, but it did not match any known species. The first complete specimen was found at another fish market, just south of Lima, Peru, in 1985. This prompted an intensive search of Peruvian fish markets, which revealed several more specimens. It was not until November 1988 that a fully adult male was found, on a deserted beach north of Lima; the taxonomy of beaked whales is so complex that a fully grown male was essential for a positive identification. The new species was officially named in 1991, after the country in which the first specimens were found. Since this account is based on sketchy data from a small number of animals, the information should be taken as tentative.

• **OTHER NAMES** Peruvian Beaked Whale, Pygmy Beaked Whale.

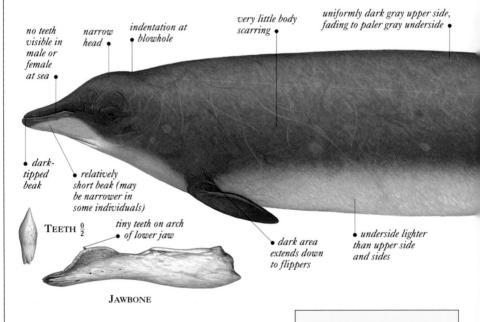

no teeth visible in male or female at sea

narrow head

indentation at blowhole

very little body scarring

uniformly dark gray upper side, fading to paler gray underside

dark-tipped beak

relatively short beak (may be narrower in some individuals)

TEETH $\frac{0}{2}$

tiny teeth on arch of lower jaw

dark area extends down to flippers

underside lighter than upper side and sides

JAWBONE

BEHAVIOR

Field identification is likely to be very difficult. Current information is based on only a handful of observations. Strandings have been of lone animals, but almost all possible sightings are of pairs (with one exception, when 2 adults and a calf were seen together). Confusion is most likely with Hector's Beaked Whale (p.128), which also occurs in pairs; nothing is known about behavioral differences. Individuals observed in 5 possible sightings in 1986 and 1988 were readily approachable. Blow inconspicuous. Appears to feed in mid to deep waters.

ID CHECKLIST

- small size
- dark, indistinct coloration
- small, triangular dorsal fin
- little body scarring
- small beak and sloping forehead
- no teeth visible
- inconspicuous blow
- may be seen in pairs
- may be approachable

Group size 2–3, information based on only a few possible sightings	Fin position Far behind center

| Status Unknown | Population Unknown | Threats |

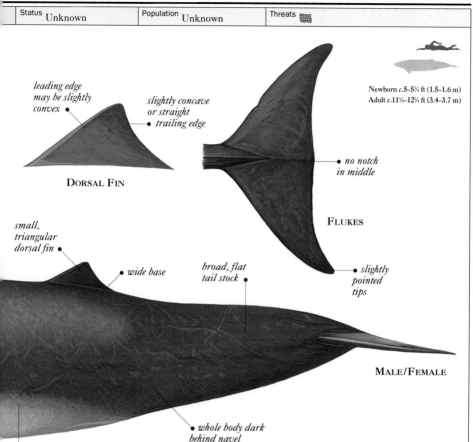

Newborn c.5–5¼ ft (1.5–1.6 m)
Adult c.11¼–12¼ ft (3.4–3.7 m)

leading edge
may be slightly
convex

slightly concave
or straight
trailing edge

no notch
in middle

DORSAL FIN

FLUKES

small,
triangular
dorsal fin

wide base

broad, flat
tail stock

slightly
pointed
tips

MALE/FEMALE

whole body dark
behind navel

spindle-
shaped body

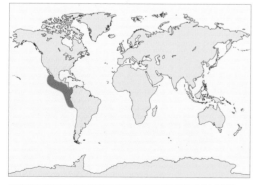

MID TO DEEP WATERS IN THE EASTERN TROPICAL PACIFIC,
PRIMARILY OFF THE COAST OF PERU

WHERE TO LOOK
Map shows broad range of strandings, incidental captures in fishing nets, and possible sightings. May be endemic to the eastern tropical Pacific, but such a limited number of records gives no clear picture of distribution and map is unlikely to show limits of range. Most strandings and incidental captures between about 11° S and 15° S along the coasts of Ica and Lima, southern and central Peru. A handful of tentative sightings have been made off the coast of central Peru. Two recent strandings (January and April 1990) in Bahía de la Paz, Baja California, Mexico, are the first outside Peruvian waters. There are no confirmed records between Peru and Baja California. Southern Peru is probably close to southern limit of its range.

| Birth wt Unknown | Adult wt Unknown | Diet |

Family ZIPHIIDAE	Species *Mesoplodon stejnegeri*	Habitat 〰〰

STEJNEGER'S BEAKED WHALE

Stejneger's Beaked Whale is inconspicuous at sea and seldom seen alive. It is probably rare, though it may simply have escaped notice in areas where there has been little research work. Females and young males have no erupted teeth and are probably impossible to distinguish from other *Mesoplodon* species. Mature males are distinctive, with 2 massive, laterally compressed teeth set about 8 in (20 cm) from the tip of the lower jaw and a strongly arched mouth line; the

teeth sometimes converge toward one another and cut into the upper "lip." Juveniles may have striking light streaks in the neck region. Confusion is likely with Hubbs' Beaked Whale (p.118) – although the white "cap" of Hubbs' should be a distinguishing feature – and is also possible with Cuvier's Beaked Whale (p.142).
• **OTHER NAMES** Saber-toothed Beaked Whale, Bering Sea Beaked Whale, North Pacific Beaked Whale.

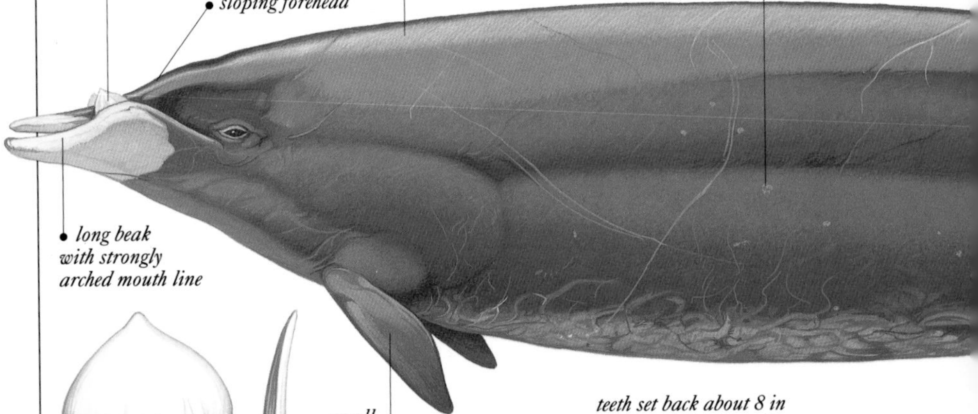

single pair of teeth erupt
• in middle of mouth line

dark, gently
• sloping forehead

black, dark gray, or
• brown upper side

white oval scars may
be visible on flanks •

• long beak
with strongly
arched mouth line

small, •
narrow
flippers

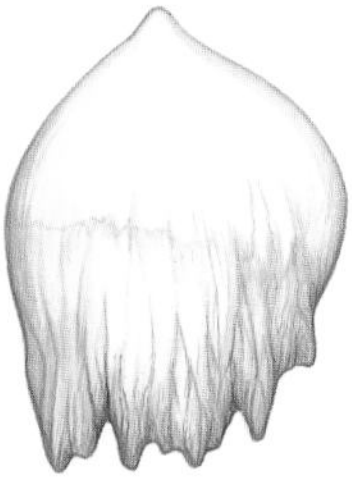

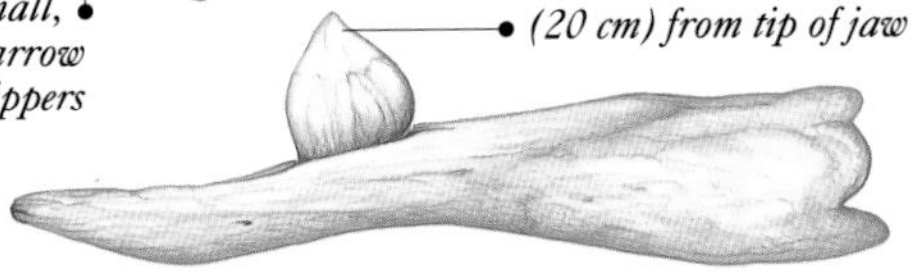

teeth set back about 8 in
• (20 cm) from tip of jaw

**MALE
JAWBONE**

SIDE  **FRONT** **TEETH $\frac{0}{2}$**

teeth do
not erupt •

BEHAVIOR

Small groups sometimes travel abreast, almost touching one another, and may surface and submerge in unison. Reports of 5 or 6 shallow dives, followed by long dives of 10 to 15 minutes. Diving involves a slow, casual roll at the surface. Groups usually include both small and large animals, suggesting a mixing of ages and/or sexes. Blow sometimes visible, but normally low and inconspicuous. Not very approachable.

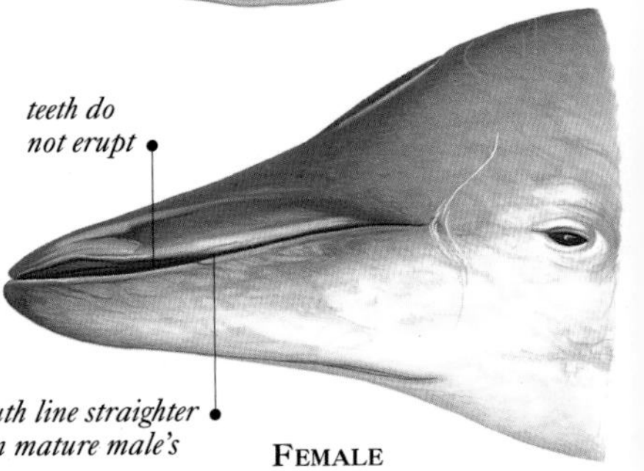

mouth line straighter •
than mature male's

FEMALE

Group size 5–15 (1–3)	Fin position Far behind center

Status Rare	Population Unknown	Threats

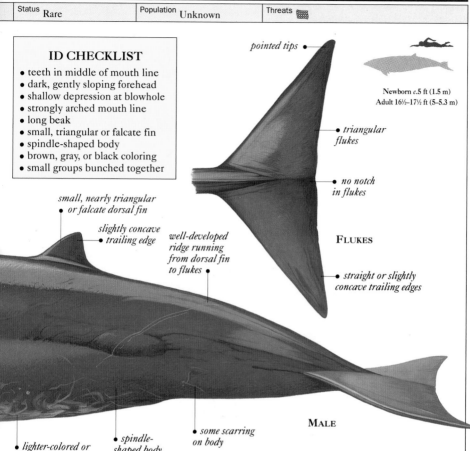

ID CHECKLIST

- teeth in middle of mouth line
- dark, gently sloping forehead
- shallow depression at blowhole
- strongly arched mouth line
- long beak
- small, triangular or falcate fin
- spindle-shaped body
- brown, gray, or black coloring
- small groups bunched together

pointed tips

Newborn *c*.5 ft (1.5 m)
Adult 16½–17½ ft (5–5.3 m)

triangular flukes

no notch in flukes

FLUKES

small, nearly triangular or falcate dorsal fin

slightly concave trailing edge

well-developed ridge running from dorsal fin to flukes

straight or slightly concave trailing edges

lighter-colored or scarred underside

spindle-shaped body

some scarring on body

MALE

WHERE TO LOOK

Distribution confused, because many earlier reports have since been re-identified as other *Mesoplodon* species. It is known mostly from strandings, but some sightings made by experienced observers. Majority of records from Alaskan waters, especially the Aleutian Islands (which appear to be the center of distribution). Although it is sometimes known as the Bering Sea Beaked Whale, seems to occur only in deep waters of the extreme southern part of the Bering Sea and probably not in shallow waters farther north. Small, possibly distinct, population also occurs in the Sea of Japan, especially off the coasts of Honshu and southern Hokkaido. Prefers deep water over the continental shelf.

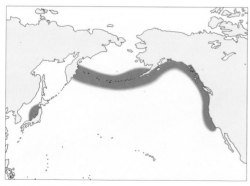

COLD TEMPERATE AND SUBARCTIC WATERS OF THE NORTH PACIFIC AND SEA OF JAPAN

Birth wt Unknown	Adult wt 1–1.5 tons	Diet

Family ZIPHIIDAE	Species *Tasmacetus shepherdi*	Habitat 〰

SHEPHERD'S BEAKED WHALE

One of the least known cetaceans, Shepherd's Beaked Whale has been recorded only about 20 times from strandings and just a few possible sightings. Increased research in the southern hemisphere may reveal many more specimens but, on current evidence, it appears to be extremely rare. There is little information on its appearance – most records are based on partly decomposed animals washed ashore – and it may be impossible to identify at sea; the most distinctive features are probably the steep rounded forehead, the long, narrow, pointed beak, and the diagonal stripes on its sides. It is the only beaked whale with a full set of functional teeth; these occur in both jaws of both sexes, although only the male has a pair of larger teeth at the tip of the lower jaw.

• **OTHER NAMES** Tasman Whale, Tasman Beaked Whale.

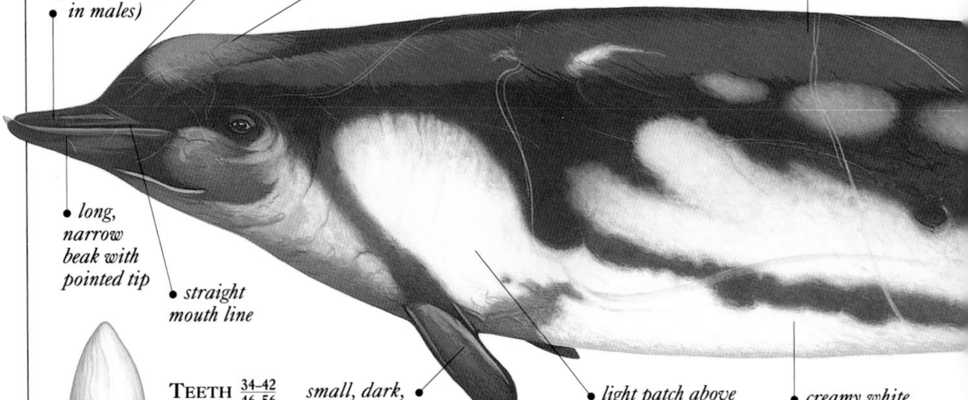

lower jaw contains 2 longer teeth at tip (erupt only in males)

steep, rounded forehead

top of head may be lighter than upper side of body

dark brownish black upper side

long, narrow beak with pointed tip

straight mouth line

TEETH $\frac{34-42}{46-56}$

small, dark, narrow flippers

light patch above flippers continuous with pale underside

creamy white underside

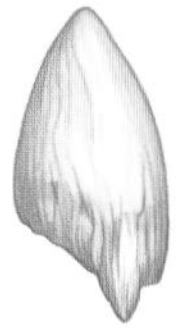

small, conical teeth occur in upper and lower jaws of both sexes

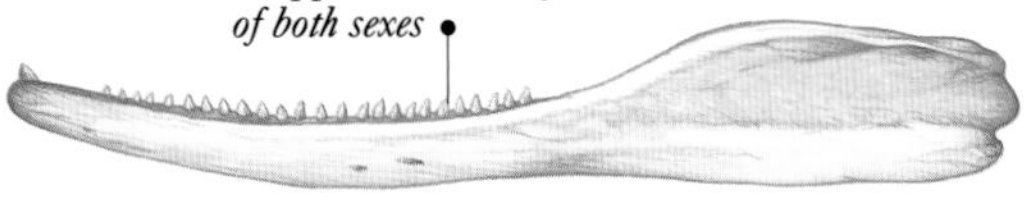

MALE JAWBONE

less complex side patterning than most other specimens

BEHAVIOR
Almost nothing known. Probably a deep diver. Limited evidence from analysis of one whale's stomach contents suggests that, unlike most other beaked whales, it eats mainly fish rather than squid (this could explain why it retains a full set of teeth). One possible sighting in New Zealand suggests an indistinct blow. Lack of sightings may be due to inconspicuous behavior, or rarity, or both.

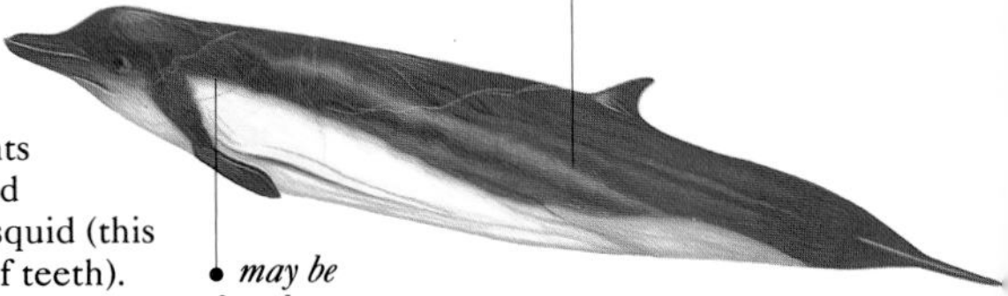

may be female or juvenile

NEW ZEALAND SPECIMEN (FOUND 1951)

Group size Unknown	Fin position Far behind center

Status Unknown	Population Unknown	Threats Unknown

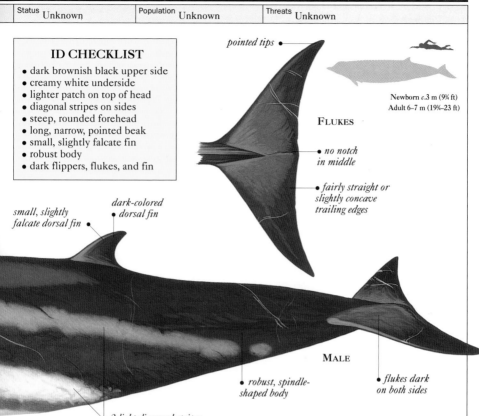

ID CHECKLIST

- dark brownish black upper side
- creamy white underside
- lighter patch on top of head
- diagonal stripes on sides
- steep, rounded forehead
- long, narrow, pointed beak
- small, slightly falcate fin
- robust body
- dark flippers, flukes, and fin

pointed tips

Newborn *c.*3 m (9¾ ft)
Adult 6–7 m (19¾–23 ft)

FLUKES

- *no notch in middle*

- *fairly straight or slightly concave trailing edges*

small, slightly falcate dorsal fin

dark-colored dorsal fin

robust, spindle-shaped body

MALE

- *flukes dark on both sides*

2 light diagonal stripes on each side (variable)

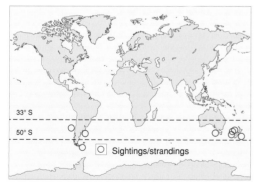

33° S
50° S

◯ Sightings/strandings

COLD TEMPERATE WATERS OF THE SOUTHERN HEMISPHERE, PREDOMINANTLY NEW ZEALAND

WHERE TO LOOK
Until 1970, all records were from New Zealand, but strandings have since been recorded in Australia, Chile, Argentina, and Tristan da Cunha in the South Atlantic. Most records between 33° S and 50° S and, even now, more than half are from New Zealand. May be circumpolar, but too little information to be certain; has not yet been recorded in South Africa but prefers cold water, so may well be sighted in the cold Benguela Current off the western coast. Too few records to make assumptions about seasonal movements. Probably lives mainly far offshore, well away from coasts; however, where there is a narrow continental shelf, may sometimes occur in deep water close to shore.

Birth wt Unknown	Adult wt *c.*2–3 tons	Diet

Family ZIPHIIDAE	Species *Ziphius cavirostris*	Habitat 〜〜

CUVIER'S BEAKED WHALE

Cuvier's Beaked Whale appears to be one of the most widespread and abundant of the beaked whales, although it is generally inconspicuous and rarely seen at sea. It is known mainly from strandings. There is so much color variation and scarring that no two look alike. It could be confused with either of the bottlenose whales (pp.108–111), but has a more gently sloping forehead and a smaller, less distinct beak. Confusion is also possible with other species

of beaked whale and the Minke Whale (p.56), but the shape of the head and beak – sometimes described as resembling a goose's beak – and the 2 small teeth at the tip of the lower jaw (in males only) should be distinctive; the teeth are sometimes covered in growths of barnacles.

• **OTHER NAMES** Cuvier's Whale, Goose-beaked Whale, Goosebeak Whale.

creamy white or white forehead, beak, and chin

indentation behind blowhole

upper side of old males can be almost white in front of dorsal fin

swirling patterns typical of many animals

indistinct beak (becomes less distinct with age)

2 small teeth just visible when mouth closed

long, white scars on sides and upper side

small flippers

TEETH $\frac{0}{2}$

heavy lower jaw, with 2 conical teeth at tip

MALE JAWBONE

COLOR VARIATIONS
The color varies according to location, sex, or age. Older animals can be so white that they may be confused with Belugas or Risso's Dolphins.

BEHAVIOR
Normally avoids boats but is occasionally inquisitive and approachable, especially around Hawaii. Breaching has been observed, though is probably rare: body rises almost vertically and comes completely out of water before falling back awkwardly. Blow directed slightly forward and to left but is low and inconspicuous; may be visible immediately after a long dive. Dives typically last 20 to 40 minutes, possibly with 2 to 3 blows 10 to 20 seconds apart in between. Animal seems to lurch through the water and may expose head when swimming fast; there is usually a good view of the dorsal fin. Arches its back steeply before a deep dive and may lift flukes above the surface. Found stranded more often than most other beaked whales.

ID CHECKLIST
• "goose beak" head shape
• short, upturned mouth line
• small head, often pale
• long, robust body
• indentation behind blowhole
• small teeth at tip of jaw
• long and circular scars
• lurches through water
• usually alone or in small groups

Group size 1–10 (1–25), lone individuals are usually old males	Fin position Far behind center

Status Unknown	Population Unknown	Threats Unknown

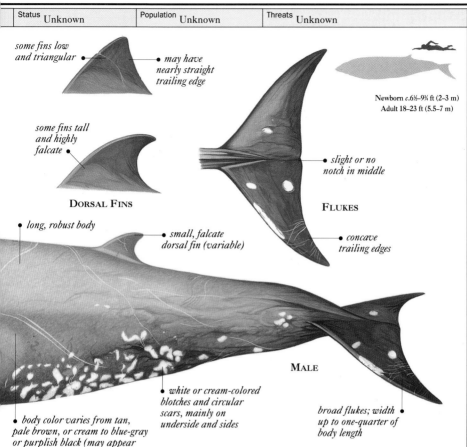

some fins low and triangular •

• *may have nearly straight trailing edge*

Newborn *c.*6½–9¾ ft (2–3 m)
Adult 18–23 ft (5.5–7 m)

some fins tall and highly falcate •

• *slight or no notch in middle*

DORSAL FINS

FLUKES

• *long, robust body*

• *small, falcate dorsal fin (variable)*

• *concave trailing edges*

MALE

• *white or cream-colored blotches and circular scars, mainly on underside and sides*

• *body color varies from tan, pale brown, or cream to blue-gray or purplish black (may appear reddish in bright sunlight)*

• *broad flukes; width up to one-quarter of body length*

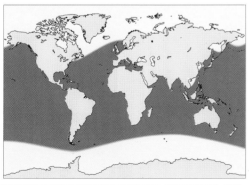

WORLDWIDE DISTRIBUTION IN TROPICAL, SUBTROPICAL, AND TEMPERATE WATERS

WHERE TO LOOK

Distribution known mainly from large number of strandings, with relatively small number of sightings. It seems to be one of the most cosmopolitan of beaked whales, with a very broad range in the Atlantic, Pacific, and Indian Oceans. Absent only from polar waters (in both hemispheres). Known around many oceanic islands, and relatively common in enclosed seas such as the Mediterranean and Sea of Japan. Resident year-round in Hawaiian waters and several other areas; no migrations are known. Rarely found close to mainland shores, except in submarine canyons or in areas where the continental shelf is narrow and coastal waters are deep.

Birth wt *c.*550 lb (250 kg)	Adult wt 2–3 tons	Diet

BLACKFISH

T HE BLACKFISH, also known as smaller, toothed whales, are generally considered more closely related to dolphins than other whale species, and are usually included in the dolphin family (the word *whale* is, broadly speaking, used to indicate size rather than zoological affinity). However, they are unlike most dolphins in appearance, and several authorities classify at least some of them separately. None of them seems to undertake regular long migrations (though they move around according to food supply and other local conditions), and most prefer deeper waters.

beak poorly defined or absent •   • *single blowhole*   *prominent dorsal fin*

CHARACTERISTICS
Blackfish are gregarious by nature, and tend to live in well-structured groups. They may also associate with a variety of other whales and dolphins. Except for the Pygmy Killer Whale and Melon-headed Whale, they are easy to approach and quite distinctive at close range. Although they share many common characteristics there are also many variations within the family; in size alone, they range from as little as 7 ft (2.1 m) long and 170 lb (110 kg) in the Pygmy Killer Whale to a maximum of 32¼ ft (9.8 m) and 9 tons in the Killer Whale.

gray-white anchor patch on chest (variable)

belly paler than upper side and sides (except False Killer Whale)

long or wide flippers

MELON-HEADED WHALE

re-enters water almost instantly

body leaves water completely

breaks surface at shallow angle

shallow dive before next leap

DIVE SEQUENCE
When traveling fast, blackfish frequently make low, shallow leaps out of the water. At other times, they have a more regular surfacing pattern.

LYING ON ONE SIDE
Pilot whales sometimes roll over onto one side and lie in the water, with one flipper and a fluke waving in the air. At a distance, they may resemble sea lions, which often snooze near the surface with their flippers out of the water in a similar way. The whale's head usually stays underwater.

DAYTIME RESTING
Pygmy Killer Whales and other blackfish probably feed mainly at night and rest, socialize, or travel near the surface during the day.

dark body color, with some white or gray markings

LONG-FINNED PILOT WHALE
The Long-finned Pilot Whale shows several physical features common to all blackfish; however, some of these can be difficult to see in the wild.

notch in middle of flukes

ANCHOR PATCH
All blackfish (except the Killer Whale) have a gray-white anchor or W-shaped patch on their chests; this varies according to the species and individual.

SPECIES IDENTIFICATION

PYGMY KILLER WHALE (p.146) *Small, shy, and little-known whale, best identified by its rounded head and dark cape.*

MELON-HEADED WHALE (p.156) *Similar to the Pygmy Killer Whale but has a more pointed beak and long, sharply pointed flippers.*

FALSE KILLER WHALE (p.158) *Acrobatic, playful whale that readily approaches boats; has unique S-shaped flippers.*

SHORT-FINNED PILOT WHALE (p.148) *Almost identical to the Long-finned Pilot Whale but has shorter flippers and fewer teeth.*

LONG-FINNED PILOT WHALE (p.150) *Often found in colder waters than the Short-finned Pilot Whale; has exceptionally long flippers.*

KILLER WHALE (p.152) *Largest of the group; has distinctive black and white body coloration and a prominent dorsal fin.*

| Family DELPHINIDAE | Species *Feresa attenuata* | Habitat 〰️ |

PYGMY KILLER WHALE

The Pygmy Killer Whale is a little-known animal and is rarely seen in the wild, though it is widely distributed and could be found almost anywhere in deep waters of the tropics or subtropics. It is similar in size to many dolphins, though it is most likely to be confused with the very similar Melon-headed Whale (p.156). There are many subtle differences, but none is easy to see in the wild except at close range. The dark cape is possibly the best distinguishing feature, but the white chin is characteristic too, although it is not present in all individuals; the shape of the head and flippers may also be useful.

Generally speaking, if a small number of animals are seen together (fewer than 50) they are more likely to be Pygmy Killer Whales. In captivity, they have been aggressive toward people and other cetaceans; they may deserve the name "killer" more than the larger Killer Whale. Evidence suggests that they prey on dolphins in the wild.
• **OTHER NAMES** Slender Blackfish, Slender Pilot Whale.

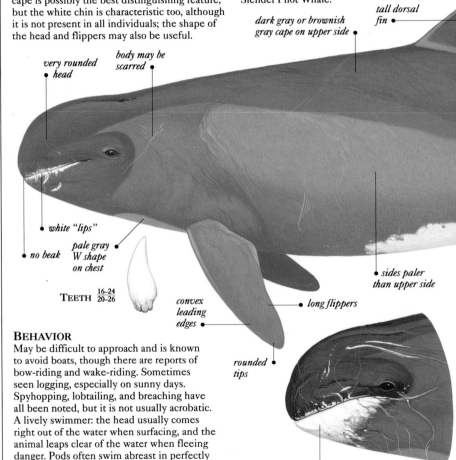

tall dorsal fin •

dark gray or brownish gray cape on upper side •

very rounded • head

body may be scarred •

white "lips" •

no beak •

pale gray W shape on chest •

TEETH $\frac{16-24}{20-26}$

convex leading edges •

• long flippers

sides paler than upper side •

rounded • tips

some individuals have a white chin •

HEAD

BEHAVIOR
May be difficult to approach and is known to avoid boats, though there are reports of bow-riding and wake-riding. Sometimes seen logging, especially on sunny days. Spyhopping, lobtailing, and breaching have all been noted, but it is not usually acrobatic. A lively swimmer: the head usually comes right out of the water when surfacing, and the animal leaps clear of the water when fleeing danger. Pods often swim abreast in perfectly coordinated "chorus lines" and, when alarmed, bunch together to rush away. Growling sounds may be heard above the surface. Often strands.

| Group size 15–25 (1–50), several hundred may be seen together (rare) | Fin position Center |

Status	Population	Threats
Unknown	Unknown	

ID CHECKLIST

- robust, dark-colored body
- dark cape
- rounded head with no beak
- pale gray sides; white underside
- some animals have a white chin
- prominent, falcate fin
- short, slightly rounded flippers
- generally elusive
- lively swimmer

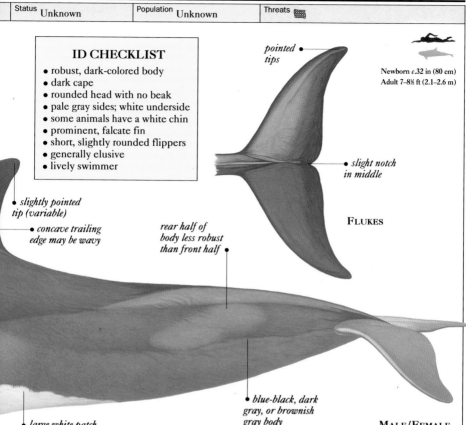

pointed tips

Newborn *c*.32 in (80 cm)
Adult 7–8½ ft (2.1–2.6 m)

slight notch in middle

FLUKES

slightly pointed tip (variable)

concave trailing edge may be wavy

rear half of body less robust than front half

blue-black, dark gray, or brownish gray body

MALE/FEMALE

large white patch on belly, split into 2 halves by deep groove

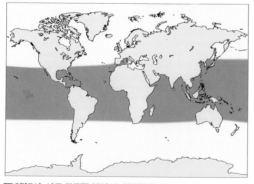

TROPICAL AND SUBTROPICAL OFFSHORE WATERS
AROUND THE WORLD

WHERE TO LOOK

Distribution is poorly known from sparse but widely distributed records worldwide. Occurs in deep, warm waters, rarely close to shore (except near oceanic islands). Mainly tropical, but occasionally strays into warm temperate regions. Seen relatively frequently in the eastern tropical Pacific, Hawaii, and Japan, though not particularly abundant anywhere; however, tends to avoid boats, so may be more common than records suggest. No migrations are known; thought to occur year-round in well-studied areas such as Sri Lanka, in the Indian Ocean, and St. Vincent, in the Caribbean.

Birth wt	Adult wt	Diet
Unknown	*c*.240–375 lb (110–170 kg)	

Family DELPHINIDAE	Species *Globicephala macrorhynchus*	Habitat ▰≈≈ ≈≈

SHORT-FINNED PILOT WHALE

The Short-finned Pilot Whale is a distinctive animal but, at sea, is virtually impossible to tell apart from its close relative, the Long-finned Pilot Whale (p.150). However, there are subtle differences between the two species: mainly the length of the flippers, the number of teeth, and the shape of the skull. Also, the Short-finned tends to prefer warmer waters. There is little overlap in range. Short-finned Pilot Whales are often found in the company of Bottlenose Dolphins and other small cetaceans, although they have been known to attack them. They are social animals, with close matrilineal associations. When traveling, pods may swim abreast in a line several miles across.

• OTHER NAMES Pothead Whale, Shortfin Pilot Whale, Pacific Pilot Whale.

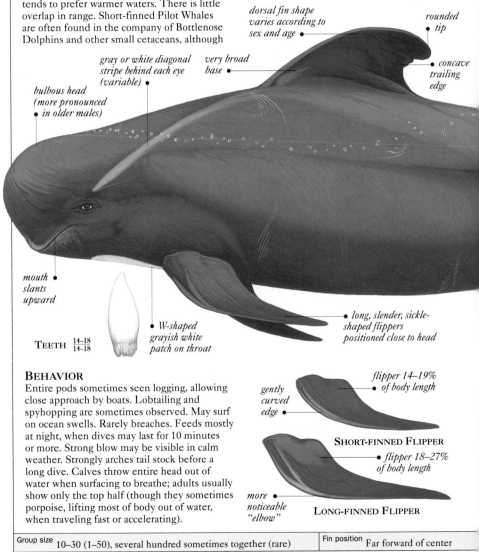

dorsal fin shape varies according to sex and age •

rounded • tip

gray or white diagonal stripe behind each eye (variable) •

very broad base •

concave trailing edge

bulbous head (more pronounced • in older males)

mouth • slants upward

TEETH $\frac{14-18}{14-18}$

• W-shaped grayish white patch on throat

• long, slender, sickle-shaped flippers positioned close to head

BEHAVIOR

Entire pods sometimes seen logging, allowing close approach by boats. Lobtailing and spyhopping are sometimes observed. May surf on ocean swells. Rarely breaches. Feeds mostly at night, when dives may last for 10 minutes or more. Strong blow may be visible in calm weather. Strongly arches tail stock before a long dive. Calves throw entire head out of water when surfacing to breathe; adults usually show only the top half (though they sometimes porpoise, lifting most of body out of water, when traveling fast or accelerating).

gently curved edge •

flipper 14–19% • of body length

SHORT-FINNED FLIPPER

• flipper 18–27% of body length

more • noticeable "elbow"

LONG-FINNED FLIPPER

Group size 10–30 (1–50), several hundred sometimes together (rare)	Fin position Far forward of center

Status Common	Population Unknown	Threats

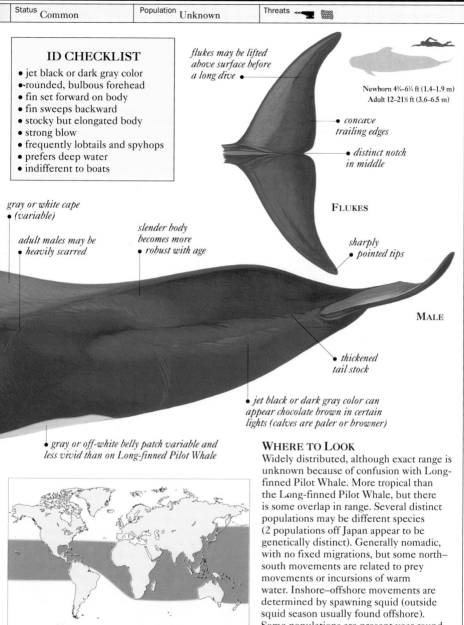

ID CHECKLIST

- jet black or dark gray color
- rounded, bulbous forehead
- fin set forward on body
- fin sweeps backward
- stocky but elongated body
- strong blow
- frequently lobtails and spyhops
- prefers deep water
- indifferent to boats

flukes may be lifted above surface before a long dive

Newborn 4¾–6¼ ft (1.4–1.9 m)
Adult 12–21½ ft (3.6–6.5 m)

concave trailing edges

distinct notch in middle

FLUKES

gray or white cape (variable)

adult males may be heavily scarred

slender body becomes more robust with age

sharply pointed tips

MALE

thickened tail stock

jet black or dark gray color can appear chocolate brown in certain lights (calves are paler or browner)

gray or off-white belly patch variable and less vivid than on Long-finned Pilot Whale

TROPICAL, SUBTROPICAL, AND WARM TEMPERATE OCEANS AROUND THE WORLD

WHERE TO LOOK

Widely distributed, although exact range is unknown because of confusion with Long-finned Pilot Whale. More tropical than the Long-finned Pilot Whale, but there is some overlap in range. Several distinct populations may be different species (2 populations off Japan appear to be genetically distinct). Generally nomadic, with no fixed migrations, but some north–south movements are related to prey movements or incursions of warm water. Inshore–offshore movements are determined by spawning squid (outside squid season usually found offshore). Some populations are present year-round, such as in Hawaii and the Canary Islands. Prefers deep water: look at the edge of the continental shelf and over deep submarine canyons. Susceptible to mass strandings.

Birth wt 135 lb (60 kg)	Adult wt 1–4 tons	Diet

Family DELPHINIDAE	Species *Globicephala melas*	Habitat

LONG-FINNED PILOT WHALE

At sea, the Long-finned Pilot Whale is almost impossible to distinguish from the Short-finned Pilot Whale (p.148). However, there are a few minor differences: the Long-finned has longer flippers and, in most cases, more teeth than the Short-finned; the shape of the skull also differs slightly. Fortunately (for identification purposes) there seems to be only a small overlap in their range. The Long-finned Pilot Whale is often found with other small cetaceans, such as Minke Whales, Common Dolphins, Bottlenose Dolphins, and Atlantic White-sided Dolphins. Its dorsal fin shape varies according to age and sex: it is sickle-shaped in young animals, relatively upright in adult females, and has a longer base and is more bulbous in adult males. The species has been heavily exploited over the centuries but is still fairly abundant.

• **OTHER NAMES** Pothead Whale, Caaing Whale, Longfin Pilot Whale, Atlantic Pilot Whale; formerly *G. melaena*.

low but prominent dorsal fin (variable)

forehead may overhang beak, especially in older males

gray or white diagonal stripe behind each eye (variable)

W-shaped grayish white patch on throat

TEETH $\frac{16-24}{16-24}$

"elbow" becomes more noticeable with age

long, slender flippers, up to one-fifth of body length, are positioned close to head

BEHAVIOR

Pods sometimes rest motionless at the surface, allowing boats to approach closely. Known to bow-ride. Lobtailing and spyhopping are often observed. Young animals may breach, but this is rare in adults. It generally takes several quick breaths and then submerges for a few minutes (feeding dives may last for 10 minutes or more). Its strong blow, more than 39 in (1 m) high, is sometimes visible in good conditions and can be heard. Capable of diving to at least 1,965 ft (600 m), but most dives are 100–195 ft (30–60 m).

ID CHECKLIST

• jet black or dark gray color
• rounded, bulbous forehead
• fin sweeps backward
• fin set forward on body
• stocky but elongated body
• exceptionally long flippers
• frequently lobtails and spyhops
• prefers deep water

Group size 10–50 (1–100), hundreds or thousands may gather	Fin position Far forward of center

Status Common	Population Unknown	Threats

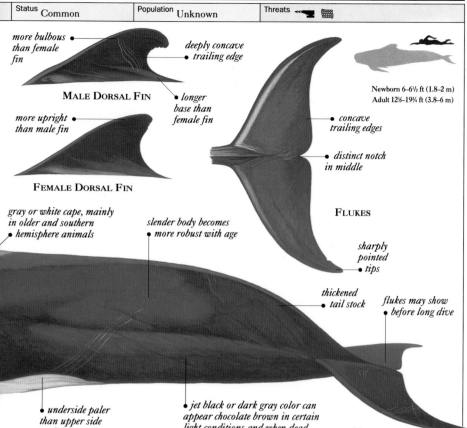

more bulbous
than female
fin

deeply concave
• trailing edge

MALE DORSAL FIN

• longer
base than
female fin

more upright •
than male fin

FEMALE DORSAL FIN

Newborn 6–6½ ft (1.8–2 m)
Adult 12½–19¾ ft (3.8–6 m)

• concave
trailing edges

• distinct notch
in middle

FLUKES

gray or white cape, mainly
in older and southern
• hemisphere animals

slender body becomes
• more robust with age

sharply
pointed
• tips

thickened
• tail stock

flukes may show
• before long dive

• underside paler
than upper side
and sides

• jet black or dark gray color can
appear chocolate brown in certain
light conditions and when dead
(calves are paler or browner)

MALE

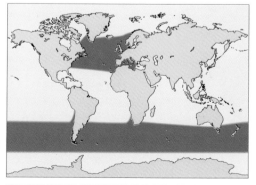

COLD TEMPERATE AND SUBPOLAR WATERS OF ALL OCEANS
EXCEPT THE NORTH PACIFIC

WHERE TO LOOK

Two distinct populations are recognized:
in the southern hemisphere (where
associated with the Humboldt, Falklands,
and Benguela Currents) and in the North
Atlantic. These are geographically
separated by the wide tropical belt and
may be different species or subspecies
(*edwardii* in the south and *melas* in the
north). Both prefer deep water. Some
live permanently offshore or inshore,
while others make inshore (summer and
autumn) to offshore (winter and spring)
migrations according to the abundance of
squid. Good place to look is over the edge
of the continental shelf. It is one of the
most commonly mass-stranded whales.

Birth wt *c.*165 lb (75 kg)	Adult wt 1.8–3.5 tons	Diet

Family DELPHINIDAE	Species *Orcinus orca*	Habitat 〰️〰️ 〰️〰️

KILLER WHALE

The Killer Whale is the largest member of the dolphin family (see p.144). Its distinctive jet black, brilliant white, and gray markings, and the huge dorsal fin of the male, make it relatively easy to identify. However, at a distance the female or juvenile may be confused with Risso's Dolphin (p.206), the False Killer Whale (p.158), or even Dall's Porpoise (p.248). Single animals may be encountered but close-knit family groups, known as pods, are typical. Two or more pods may come together temporarily to form superpods, which may contain 150 or more whales. Members of a pod usually stay together for life, and groups of closely related pods (known as clans) develop their own unique dialects. Despite their name, Killer Whales do not harm people in the wild, and aggression within a pod is rare.
• **OTHER NAMES** Orca, Great Killer Whale, Grampus.

very tall dorsal fin, up to 6 ft (1.8 m), especially in older males

dorsal fin may lean forward

gray saddle patch (variable)

conspicuous, elliptical white patch behind each eye

rounded head tapers to point

white chin

white chest

BEHAVIOR

Inquisitive and approachable. Rarely bow-rides or wake-rides, but breaching, lobtailing, flipper-slapping, and spy-hopping are often observed. Other behavior includes beach-rubbing; speed-swimming, when most of body leaves water as it surfaces to breathe; logging, when whole pod faces in same direction; and occasionally dorsal fin slapping, involving a sudden roll onto one side to slap fin onto water's surface. Can travel at up to 34 mph (55 km/h). Blow often visible in cool air, when it is low and bushy.

large, paddle-shaped flippers grow with age; may measure up to one-fifth of body length in older males

TEETH
20–26
20–26

ID CHECKLIST

• black and white coloration
• white patch behind each eye
• gray saddle patch
• large, paddle-shaped flippers
• robust, heavy body and tall fin
• pronounced sexual differences
• acrobatic and active at surface
• lives in mixed family groups

Group size 3–25 (1–50), several pods may meet at social gatherings	Fin position Slightly forward of center

Status Locally common	Population Unknown	Threats 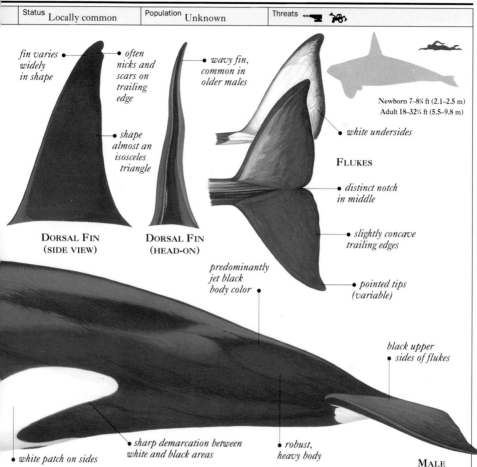

fin varies widely in shape

often nicks and scars on trailing edge

wavy fin, common in older males

Newborn 7–8¼ ft (2.1–2.5 m)
Adult 18–32¼ ft (5.5–9.8 m)

white undersides

shape almost an isosceles triangle

FLUKES

distinct notch in middle

DORSAL FIN
(SIDE VIEW)

DORSAL FIN
(HEAD-ON)

slightly concave trailing edges

predominantly jet black body color

pointed tips (variable)

black upper sides of flukes

sharp demarcation between white and black areas

robust, heavy body

white patch on sides

MALE

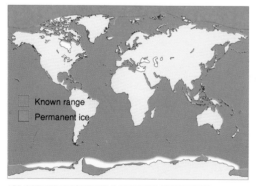

Known range

Permanent ice

ALL OCEANS OF THE WORLD, PARTICULARLY IN THE POLAR REGIONS

WHERE TO LOOK

One of the most wide-ranging mammals on earth, though distribution is patchy. Seen more often in cooler waters (especially polar regions) than in the tropics and subtropics. Sightings range from surf zone to open sea, though usually within 500 miles (800 km) of shoreline. Large concentrations may be found over continental shelf. Generally prefers deep water but is often found in shallow bays, inland seas, and estuaries (but rarely in rivers). Readily enters areas of floe ice in search of prey. No regular long migrations, but some local movements according to ice cover in high latitudes and food availability elsewhere. Stranding is rare and usually involves males.

Birth wt 395 lb (180 kg)	Adult wt 2.6–9 tons	Diet

Family DELPHINIDAE	Species *Orcinus orca*	Habitat 〰〰 〰〰

TRANSIENTS AND RESIDENTS

Studies in northwest North America suggest that there are 2 genetically distinct forms of Killer Whale, known as "transients" and "residents." With experience, these can be distinguished by differences in appearance and behavior. Transients tend to form smaller pods (1 to 7), roam over a wider area, feed predominantly on mammals, vocalize less frequently, make abrupt changes in swimming direction, and often stay underwater for 5 to 15 minutes at a time. They also have more pointed, centrally positioned dorsal fins than residents. Residents tend to form larger pods (typically 5 to 25), have smaller home ranges (at least in the summer), feed mainly on fish, vocalize frequently, keep to predictable routes, and rarely stay underwater for more than 4 minutes at a time.

dorsal fin of both sexes of calf similar to adult • female fin

smaller, more curved dorsal fin than male: up to 35 in (90 cm) long •

saddle patch less distinct or • absent

CALF

SEXUAL DIFFERENCES

In Killer Whales, there are marked differences between the sexes. Males are longer and bulkier than females: the average male length is 24 ft (7.3 m); the average length for a female is 20¼ ft (6.2 m). There is also a great difference in the size and shape of their dorsal fins.

• smaller flippers than male

FOOD AND FEEDING

The Killer Whale is a versatile predator and has one of the most varied diets of all cetaceans. It is known to eat anything from squid, fish, and birds to sea turtles, seals, and dolphins; it will even tackle animals as large as Blue Whales. Pods often cooperate during a hunt. Relationship with its prey is complex: pods tend to specialize and frequently ignore potential prey; several species of whales and dolphins will associate with Killer Whales, apparently without fear, and seem to know instinctively when there is no danger of being attacked.

teeth curve back toward throat •

broad jaw with few, large • teeth

• teeth interlock when jaws close

SKULL

Group size 3–25 (1–50), several pods may meet at social gatherings	Fin position Slightly forward of center

Status Locally common	Population Unknown	Threats

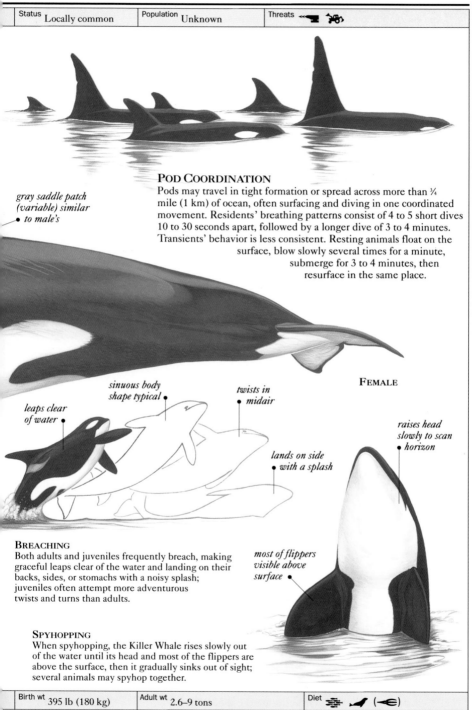

*gray saddle patch
(variable) similar
• to male's*

POD COORDINATION

Pods may travel in tight formation or spread across more than ¾ mile (1 km) of ocean, often surfacing and diving in one coordinated movement. Residents' breathing patterns consist of 4 to 5 short dives 10 to 30 seconds apart, followed by a longer dive of 3 to 4 minutes. Transients' behavior is less consistent. Resting animals float on the surface, blow slowly several times for a minute, submerge for 3 to 4 minutes, then resurface in the same place.

FEMALE

*sinuous body
shape typical •*

*twists in
• midair*

*leaps clear
of water •*

*lands on side
• with a splash*

*raises head
slowly to scan
• horizon*

BREACHING

Both adults and juveniles frequently breach, making graceful leaps clear of the water and landing on their backs, sides, or stomachs with a noisy splash; juveniles often attempt more adventurous twists and turns than adults.

*most of flippers
visible above
surface •*

SPYHOPPING

When spyhopping, the Killer Whale rises slowly out of the water until its head and most of the flippers are above the surface, then it gradually sinks out of sight; several animals may spyhop together.

Birth wt 395 lb (180 kg)	Adult wt 2.6–9 tons	Diet

Family DELPHINIDAE	Species *Peponocephala electra*	Habitat 〰️〰️〰️

MELON-HEADED WHALE

Little is known about the Melon-headed Whale, though its distribution is widespread, and it could be found almost anywhere in deep waters of the tropics or subtropics. It may associate with Fraser's Dolphins and sometimes other cetaceans such as spinner dolphins and spotted dolphins. Confusion is most likely with the Pygmy Killer Whale (p.146), which is very similar in appearance. There are subtle differences between the two species, but these are difficult to recognize except at close range. The main features to look for in the Melon-

headed Whale are the pointed, melon-shaped head, the slimmer body, and the long, sharply pointed flippers. Pygmy Killer Whales have a larger and more prominent light patch on the underside and a more distinctive cape, and some have a white chin. If the Melon-headed Whale is seen from above, its head is sharply pointed or triangular, whereas the Pygmy Killer Whale's is very rounded.

• **OTHER NAMES** Melonhead Whale, Many-toothed Blackfish, Little Killer Whale, Electra Dolphin.

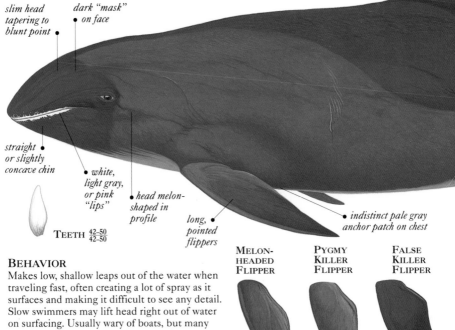

slim head tapering to blunt point •

dark "mask" • on face

straight • or slightly concave chin

• white, light gray, or pink "lips"

• head melon-shaped in profile

long, • pointed flippers

• indistinct pale gray anchor patch on chest

TEETH $\frac{42-50}{42-50}$

BEHAVIOR

Makes low, shallow leaps out of the water when traveling fast, often creating a lot of spray as it surfaces and making it difficult to see any detail. Slow swimmers may lift head right out of water on surfacing. Usually wary of boats, but many observations are in areas where tuna boats regularly chase dolphins, so behavior may differ elsewhere. Known to bow-ride for short periods and breaching has occasionally been recorded. Sometimes spyhops. Tail stock arches strongly when diving. Highly gregarious and more likely to be seen in large pods than the Pygmy Killer Whale. Animals in a pod are often tightly packed and make frequent course changes. Known to strand in large numbers.

MELON-HEADED FLIPPER

PYGMY KILLER FLIPPER

FALSE KILLER FLIPPER

one-fifth of body length; long, sharply pointed tip

one-eighth of body length; stubby, rounded tip

one-tenth of body length; S-shaped with "elbow"

Group size 100–500 (50–1,500), as many as 2,000 at some gatherings (rare)	Fin position Center

Status Unknown	Population Unknown	Threats 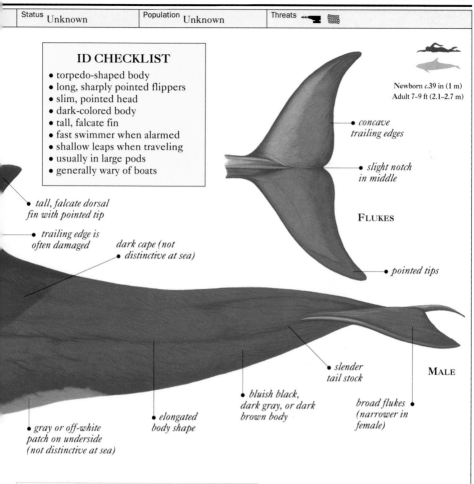

ID CHECKLIST
- torpedo-shaped body
- long, sharply pointed flippers
- slim, pointed head
- dark-colored body
- tall, falcate fin
- fast swimmer when alarmed
- shallow leaps when traveling
- usually in large pods
- generally wary of boats

Newborn *c.*39 in (1 m)
Adult 7–9 ft (2.1–2.7 m)

*• concave
trailing edges*

*• slight notch
in middle*

FLUKES

• pointed tips

*• tall, falcate dorsal
fin with pointed tip*

*• trailing edge is
often damaged*

*dark cape (not
• distinctive at sea)*

*• slender
tail stock*

MALE

*• bluish black,
dark gray, or dark
brown body*

*broad flukes •
(narrower in
female)*

*• elongated
body shape*

*• gray or off-white
patch on underside
(not distinctive at sea)*

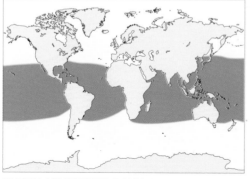

TROPICAL AND SUBTROPICAL OFFSHORE WATERS AROUND
THE WORLD

WHERE TO LOOK
Distribution thought to be continuous
across deep waters in tropics and
subtropics, though few records. However,
appears to be relatively common in the
Philippines (especially around Cebu
Island) and along the east coast of
Australia, and occurs year-round in Hawaii.
Most sightings are from the continental
shelf seaward, and around oceanic islands.
Rarely found in warm temperate waters
(northern extremes of range likely to be
associated with warm currents) and seldom
ventures close to land. No migrations are
known, and are unlikely. May be more
common than the sparse records indicate.

Birth wt Unknown	Adult wt *c.*355 lb (160 kg)	Diet

Family DELPHINIDAE	Species *Pseudorca crassidens*	Habitat 〰️ (🌊)

FALSE KILLER WHALE

The False Killer Whale seems to be fairly uncommon, but it is widely distributed and readily approaches boats. It is an exceptionally active and playful animal, especially considering its large size. In captivity, it shows less aggression than its smaller relative, the Pygmy Killer Whale, though in the wild it is believed to eat dolphins and has been seen attacking a Humpback Whale calf. However, it sometimes associates with Bottlenose Dolphins and other small cetaceans. Most pods are relatively small, although several hundred animals have been seen traveling together. Its size distinguishes it from the Pygmy Killer Whale (p.146) and Melon-headed Whale (p.156), and it is slimmer and darker than the female Killer Whale (p.154), which it also resembles. At a distance, it may be confused with a pilot whale (pp.148–151); however, look out for its slender head and body, dolphinlike dorsal fin, and more energetic behavior.

• **OTHER NAMES** False Pilot Whale, Pseudorca.

pointed tip (variable)

long, slender head tapers to rounded beak

dark head may look pale gray in certain lights

gray or off-white W shape on chest (variable)

unique "elbow" on S-shaped leading edges

short, narrow flippers set far forward on body

may have rounded tip

pointed tips

TEETH $\frac{16-22}{16-22}$

DORSAL FIN

BEHAVIOR

Fast, active swimmer. Often lifts entire head and much of body out of the water when it surfaces, and sometimes even the flippers are visible. Frequently emerges with mouth open, revealing rows of teeth. May make sudden stops or sharp turns, especially when feeding. Approaches boats to investigate, bow-ride, or wake-ride. Often breaches, usually twisting to fall back into the water on its side, causing a huge splash for a whale of its size. Makes graceful leaps clear of the water when excited, and lobtails. Seems to be susceptible to stranding, sometimes in huge numbers (more than 800 in one exceptional case).

teeth along most of length of jaw bones

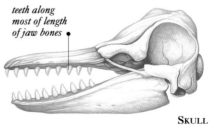

SKULL

Group size 10–50 (1–300), may be several hundred at social gatherings	Fin position Center

Status Rare	Population Unknown	Threats

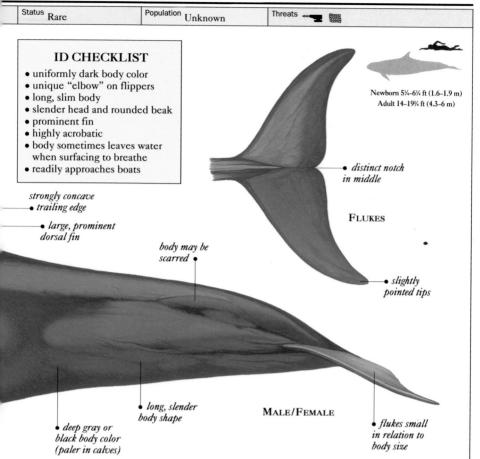

ID CHECKLIST

- uniformly dark body color
- unique "elbow" on flippers
- long, slim body
- slender head and rounded beak
- prominent fin
- highly acrobatic
- body sometimes leaves water when surfacing to breathe
- readily approaches boats

Newborn 5¼–6¼ ft (1.6–1.9 m)
Adult 14–19¾ ft (4.3–6 m)

• distinct notch in middle

FLUKES

strongly concave
• trailing edge

• large, prominent dorsal fin

body may be scarred •

• slightly pointed tips

• long, slender body shape

MALE/FEMALE

• deep gray or black body color (paler in calves)

• flukes small in relation to body size

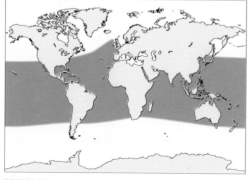

DEEP TROPICAL, SUBTROPICAL, AND WARM TEMPERATE WATERS, MAINLY OFFSHORE

WHERE TO LOOK

Widely distributed, though not really abundant anywhere. Mainly seen in deep, offshore waters (and some semi-enclosed seas such as the Red Sea and the Mediterranean) and sometimes in deep coastal waters. Seems to prefer warmer waters and, although no fixed migrations are known, may move from north to south according to seasonal warming and cooling of the sea. There are numerous records of animals seen in cool temperate waters, although these appear to be outside the normal range. Wanderers have been recorded as far afield as Norway and Alaska.

Birth wt 175 lb (80 kg)	Adult wt 1.1–2.2 tons	Diet

OCEANIC DOLPHINS
WITH PROMINENT BEAKS

T HE DELPHINIDAE is the largest and most diverse family of cetaceans: it contains 26 currently recognized species of dolphin (pp.160–223), as well as 6 toothed whales (pp.144–159). The family includes all the oceanic, or "true," dolphins and also some coastal and partly riverine species. For the purposes of this book, the oceanic dolphins have been divided into 2 equal groups: species with prominent beaks (described here) and those without prominent beaks (pp.194–223); this is not a recognized basis for classification but is merely intended to help with identification.

CHARACTERISTICS
The 13 oceanic dolphins described in this section tend to have long, well-defined beaks, streamlined or slightly robust bodies, smoothly sloping foreheads, and a single notch in the middle of their flukes. All but 2 species (Northern and Southern Rightwhale Dolphins) also have a prominent dorsal fin, located at the center of the body; however, the fin shape varies considerably between species and, indeed, between individuals. The dolphins range in length from about 4¼ ft (1.3 m) to 12¾ ft (3.9 m).

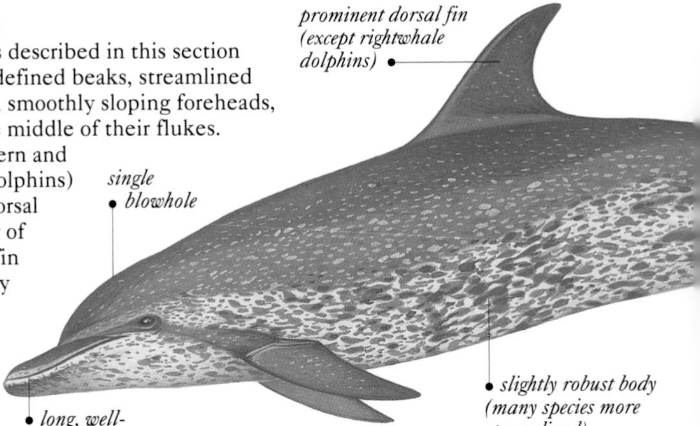

prominent dorsal fin (except rightwhale dolphins) •

single • blowhole

• slightly robust body (many species more streamlined)

• long, well-defined beak

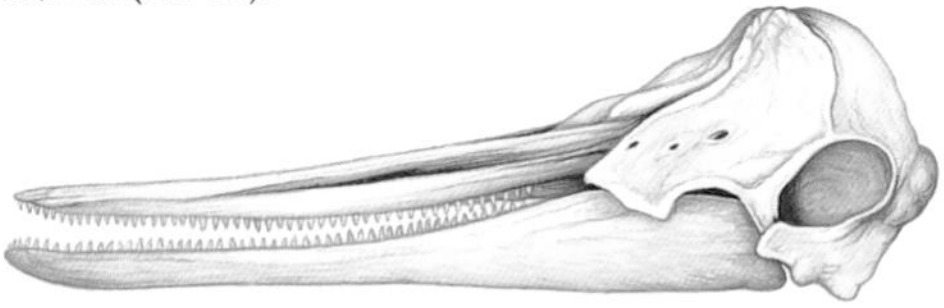

SKULL
Oceanic dolphins with prominent beaks have elongated skulls with as many as 130 conical teeth in each jaw, depending on the species. Porpoises have shorter jaws than dolphins and spade-shaped teeth.

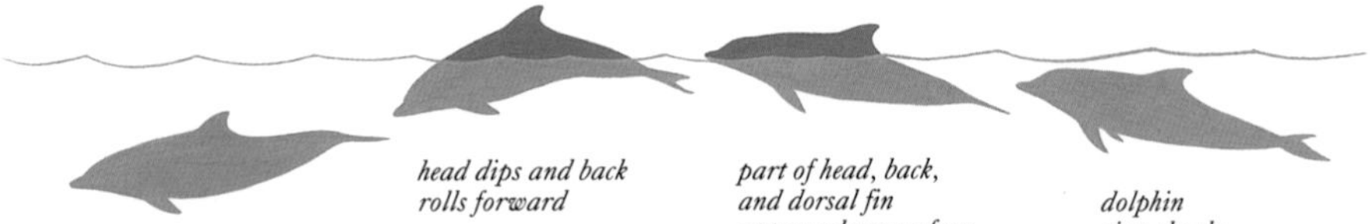

head dips and back rolls forward

part of head, back, and dorsal fin appear above surface

dolphin rises slowly

dolphin disappears; flukes may be visible when diving

BOTTLENOSE DOLPHIN

DIVE SEQUENCE (SLOW-SWIMMING)
All dolphins swim slowly at times, but coastal dolphins, which tend to feed on slower-moving prey than their offshore relatives, are rarely able to swim at extremely high speeds. Slow swimmers normally show relatively little of themselves when they surface to breathe.

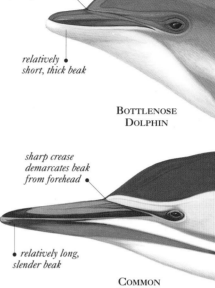

sharp crease
demarcates beak
from forehead •

relatively •
short, thick beak

**BOTTLENOSE
DOLPHIN**

NORTHERN RIGHTWHALE DOLPHIN
*Northern and Southern Rightwhale Dolphins are
exceptional, as they are the only members of the
family Delphinidae that do not have a dorsal fin.*

notch in middle
• of flukes

sharp crease
demarcates beak
from forehead •

relatively long,
• slender beak

**COMMON
DOLPHIN**

ATLANTIC SPOTTED DOLPHIN
*Members of the family Delphinidae, especially
those with prominent beaks, are the cetaceans
most people associate with the word "dolphin."
The Atlantic Spotted Dolphin is distinctive for
its extensive spotting, but illustrates many of the
features common to most members of this group.*

BEAKS
Every member of this group has a well-defined
beak, with a distinct crease at the base of the
forehead. The beak length and width vary
greatly between species. The Southern
Rightwhale Dolphin has a noticeably shorter
beak than any other species in the group but
is included here for direct comparison with
the Northern Rightwhale Dolphin.

**BOTTLENOSE
DOLPHIN**

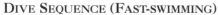

entire body comes
out of water

begins to leave water
at shallow angle

re-enters water
head first

dolphin rises
at high speed

DIVE SEQUENCE (FAST-SWIMMING)
Some dolphins are able to achieve high
swimming speeds by leaping from the water in
a series of arcs every time they need to breathe,
instead of swimming along at the surface. This
is known as "porpoising." In exceptional cases,
they can reach speeds of up to 25 mph (40 km/h).

SPECIES IDENTIFICATION

TUCUXI (p.172) *One of the smallest of all cetaceans, found in shallow coastal waters and rivers; there is great color variation between individuals and populations.*

SHORT-SNOUTED SPINNER DOLPHIN (p.180) *Has a dark gray or black dorsal cape and shorter, slightly stubbier beak than the Long-snouted Spinner Dolphin; sometimes spins longitudinally when breaching, though rarely leaps high.*

LONG-SNOUTED SPINNER DOLPHIN (p.182) *One of the most acrobatic of all cetaceans and well known for its spectacular aerial displays; there are many different varieties of this species.*

ATLANTIC HUMP-BACKED DOLPHIN (p.176) *Very similar to the Indo-Pacific Hump-backed Dolphin, but their ranges do not overlap; named after the elongated hump in the middle of its back.*

PANTROPICAL SPOTTED DOLPHIN (p.184) *Varies greatly in size, shape, and color, but most animals can be identified by their distinctive spotting and extremely active behavior.*

ATLANTIC SPOTTED DOLPHIN (p.186) *Closely resembles the Pantropical Spotted Dolphin, though has a distinctive light-colored blaze on each shoulder and spots on the underside do not merge.*

SOUTHERN RIGHTWHALE DOLPHIN (p.170) *Easily identified at sea by its very striking black and white body pattern; only dolphin in the southern hemisphere without a dorsal fin.*

SPECIES IDENTIFICATION

COMMON DOLPHIN (p.164) *Easily recognized by the hourglass pattern and tan or yellowish patch on each side; one of the more gregarious of the cetaceans.*

STRIPED DOLPHIN (p.178) *Probably one of the most common of all cetaceans, with distinctive striping and often with a bright pink underside.*

ROUGH-TOOTHED DOLPHIN (p.190) *Very distinctive dolphin, with a uniquely shaped head; however, rarely seen in the wild and poorly known.*

INDO-PACIFIC HUMP-BACKED DOLPHIN (p.174) *Very similar to the Atlantic Hump-backed Dolphin, with an elongated hump on its back; usually hard to approach.*

NORTHERN RIGHTWHALE DOLPHIN (p.168) *Unmistakable at sea, with its distinctive black upper side and sides and no dorsal fin.*

BOTTLENOSE DOLPHIN (p.192) *A highly active and well-known dolphin, with subdued gray coloring and a prominent dorsal fin; great variation in size and appearance.*

Family DELPHINIDAE	Species *Delphinus delphis*	Habitat

COMMON DOLPHIN

The Common Dolphin varies so much in appearance that more than 20 species have been proposed over the years. Only a single species is currently recognized, although there are 2 distinct forms – the short-beaked and the long-beaked – which may soon be granted species status. Confusion is most likely with Striped Dolphins (p.178), spinner dolphins (pp.180–183), spotted dolphins (pp.184–189), and Atlantic White-sided Dolphins (p.210), but the Common Dolphin's elaborate criss-cross or hourglass pattern is a good distinguishing feature. There are only subtle differences between the coloration of the sexes. Although there is some evidence of population declines in the Black Sea, the Mediterranean, and the eastern tropical Pacific, it is still one of the most abundant of all cetaceans, probably numbered in millions.
• **OTHER NAMES** Saddleback Dolphin, White-bellied Porpoise, Criss-cross Dolphin, Hourglass Dolphin, Cape Dolphin.

gray, black, purplish black, or brownish cape, with V shape under dorsal fin

tan or yellowish patch on both sides (variable)

complex coloration on head

distinct crease between beak and smoothly sloping forehead

• *gray or black beak may be white-tipped*

• *prominent beak (variable)*

• *dark circle around each eye*

dark streak from flipper to middle of lower jaw

convex leading edges

slightly pointed tips

• *broad, black or gray flippers*

TEETH $\frac{80–120}{80–120}$

BEHAVIOR
Often found in large, active schools: jumping and splashing can be seen and even heard from a considerable distance. Several members of a group often surface together. School size varies seasonally and according to the time of day. Animals bunch tightly together when frightened. Fast swimmer and energetic acrobat. Frequently porpoises, slaps water with chin, flipper-slaps, lobtails, bow-rides, and breaches (sometimes turning somersaults). Highly vocal: its high-pitched squealing can sometimes be heard above the surface. Dives can last up to 8 minutes, but usually 10 seconds to 2 minutes. May associate with other dolphins at good feeding grounds and, in the eastern tropical Pacific, with Yellowfin Tuna.

ID CHECKLIST
• dark cape with "V" under fin
• hourglass pattern on sides
• white underside and lower sides
• predominantly dark flippers, flukes, and fin
• yellowish patch on sides
• dark line from flipper to beak
• prominent fin and beak
• highly active

Group size 10–500 (1–2,000), largest groups in eastern tropical Pacific	Fin position Center

Status Common	Population Unknown	Threats

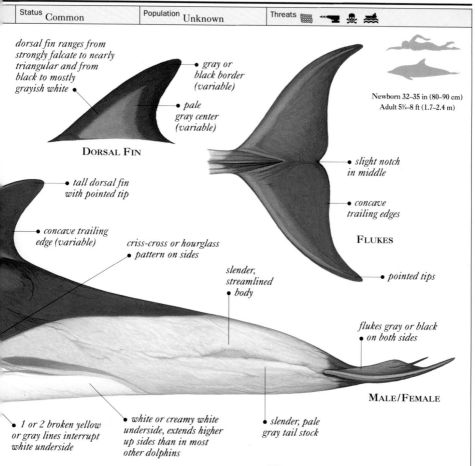

*dorsal fin ranges from
strongly falcate to nearly
triangular and from
black to mostly
grayish white*

• *gray or
black border
(variable)*

• *pale
gray center
(variable)*

DORSAL FIN

Newborn 32–35 in (80–90 cm)
Adult 5¾–8 ft (1.7–2.4 m)

• *tall dorsal fin
with pointed tip*

• *concave trailing
edge (variable)*

*criss-cross or hourglass
• pattern on sides*

*slender,
streamlined
• body*

• *slight notch
in middle*

• *concave
trailing edges*

FLUKES

• *pointed tips*

*flukes gray or black
• on both sides*

MALE/FEMALE

• *1 or 2 broken yellow
or gray lines interrupt
white underside*

• *white or creamy white
underside, extends higher
up sides than in most
other dolphins*

• *slender, pale
gray tail stock*

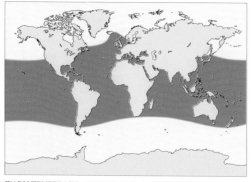

WARM TEMPERATE, SUBTROPICAL, AND TROPICAL
WATERS WORLDWIDE

WHERE TO LOOK

Widely distributed, though appear to be
many different populations. Found in
many enclosed waters such as the Red
Sea and the Mediterranean. May be less
common in the Indian Ocean. Present
all year round in some areas, but many
populations appear to move seasonally and
show local peaks of abundance at different
times of year. Usually found where surface
water temperature is 50°–82° F (10°–28° C),
limiting distribution to north and south
of range, but may follow warm water
currents beyond the normal range. Less
commonly seen in water shallower than
590 ft (180 m). Occurs over the continental
shelf, particularly in areas with high sea
floor relief, but mainly offshore.

Birth wt Unknown	Adult wt 155–245 lb (70–110 kg)	Diet

Family DELPHINIDAE	Species *Delphinus delphis*	Habitat 〜〜 (〜〜)

GEOGRAPHIC FORMS

Taxonomy of the Common Dolphin is very complicated, as there are so many variations. Research in California and Mexico has revealed 2 distinct forms: the long-beaked and the short-beaked. These show many subtle physical and behavioral differences, and recent evidence, based on morphological and genetic studies, suggests that they may be separate species. From limited observations elsewhere, these forms also appear to be distinguishable in other parts of the world. Both long-beaked and short-beaked forms also have a wide range of more subtle variations within their own populations.

These probably represent distinct races, and are not sufficiently different to grant the animals species status. Races vary mainly in body size – from an average of 6 ft (1.8 m) long in the Black Sea to 8 ft (2.4 m) in the Indian Ocean – and coloring (though most still have an instantly recognizable hourglass pattern on their sides). One well-studied variation of the long-beaked form is the so-called Baja neritic race: found in the Gulf of California (Sea of Cortez), Mexico, and the eastern tropical Pacific, north of 20° N, this form occurs mainly in shallow waters, 65–590 ft (20–180 m) deep.

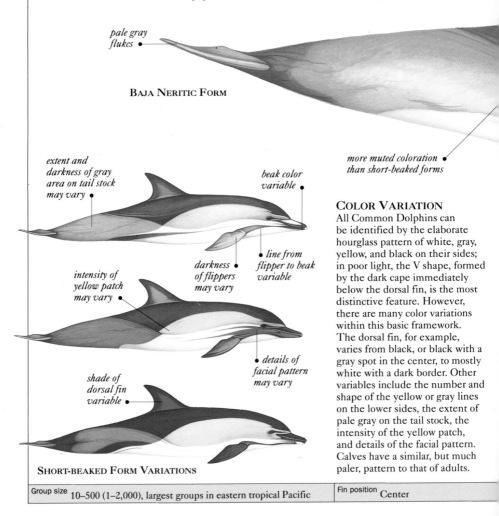

pale gray flukes •

BAJA NERITIC FORM

extent and darkness of gray area on tail stock may vary •

beak color variable •

more muted coloration • than short-beaked forms

intensity of yellow patch may vary •

darkness • of flippers may vary

• line from flipper to beak variable

shade of dorsal fin variable •

• details of facial pattern may vary

SHORT-BEAKED FORM VARIATIONS

COLOR VARIATION

All Common Dolphins can be identified by the elaborate hourglass pattern of white, gray, yellow, and black on their sides; in poor light, the V shape, formed by the dark cape immediately below the dorsal fin, is the most distinctive feature. However, there are many color variations within this basic framework. The dorsal fin, for example, varies from black, or black with a gray spot in the center, to mostly white with a dark border. Other variables include the number and shape of the yellow or gray lines on the lower sides, the extent of pale gray on the tail stock, the intensity of the yellow patch, and details of the facial pattern. Calves have a similar, but much paler, pattern to that of adults.

Group size 10–500 (1–2,000), largest groups in eastern tropical Pacific	Fin position Center

Status Common	Population Unknown	Threats 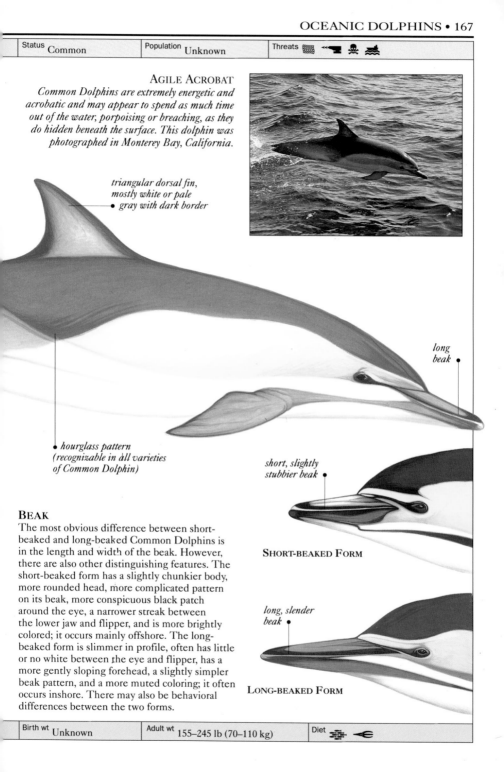

AGILE ACROBAT
*Common Dolphins are extremely energetic and
acrobatic and may appear to spend as much time
out of the water, porpoising or breaching, as they
do hidden beneath the surface. This dolphin was
photographed in Monterey Bay, California.*

triangular dorsal fin,
mostly white or pale
• gray with dark border

long
beak •

• hourglass pattern
(recognizable in all varieties
of Common Dolphin)

short, slightly
stubbier beak •

BEAK
The most obvious difference between short-
beaked and long-beaked Common Dolphins is
in the length and width of the beak. However,
there are also other distinguishing features. The
short-beaked form has a slightly chunkier body,
more rounded head, more complicated pattern
on its beak, more conspicuous black patch
around the eye, a narrower streak between
the lower jaw and flipper, and is more brightly
colored; it occurs mainly offshore. The long-
beaked form is slimmer in profile, often has little
or no white between the eye and flipper, has a
more gently sloping forehead, a slightly simpler
beak pattern, and a more muted coloring; it often
occurs inshore. There may also be behavioral
differences between the two forms.

SHORT-BEAKED FORM

long, slender
beak •

LONG-BEAKED FORM

Birth wt Unknown	Adult wt 155–245 lb (70–110 kg)	Diet

Family DELPHINIDAE	Species *Lissodelphis borealis*	Habitat 〰〰 (〰〰)

NORTHERN RIGHTWHALE DOLPHIN

This is the only dolphin in the North Pacific without a dorsal fin; consequently, it is not likely to be confused with any other cetacean. Its striking black and white body pattern is also distinctive, though in the water the body often appears to be entirely black. However, since it is probably the sleekest of all cetaceans, and because it often travels with smooth, low leaps, it could be mistaken for a sea lion or fur seal. It is very similar in appearance to its southern counterpart, the Southern Rightwhale Dolphin (p.170) but is slightly longer and usually has a less extensive area of white on its body; there is no overlap in range. It was named after the much larger Northern and Southern Right Whales (p.44), which also have finless backs. Animals with slight color variations have been found near Japan and may be a different subspecies. Calves are grayish brown or cream, developing the adult coloration during their first year. The Northern Rightwhale Dolphin often associates with Risso's Dolphins, Pacific White-sided Dolphins, and Short-finned Pilot Whales, as well as other species.

• **OTHER NAME** Pacific Rightwhale Porpoise.

no dorsal fin •

beak clearly demarcated from forehead by groove

gently sloping forehead

lower jaw extends beyond tip of upper jaw

short, slender beak

white area behind tip of lower jaw

narrow head

small, slender flippers

white chest patch, with connecting line to flukes (variable)

pointed tips

TEETH $\frac{74-98}{74-98}$

body leaves water in low-angle leaps

BEHAVIOR
When traveling fast and leaping, overall impression is of a bouncing motion; each leap can be up to 23 ft (7 m) long. Easily startled. When fleeing, group typically gathers in tight formation, with many animals leaping simultaneously, and often working the sea into a froth. It may also swim slowly, causing little disturbance of the water and exposing little of itself at the surface. Breaching, belly-flopping, side-slapping, and lobtailing fairly common. May bow-ride, but usually avoids boats.

PORPOISING
When startled, or simply traveling fast, a group of Northern Rightwhale Dolphins may make long, low leaps; they usually re-enter the water cleanly and gracefully, but sometimes belly flop or side-slap if they are fleeing danger.

Group size 5–200 (1–200), up to 3,000 have been observed together	Fin position No fin

Status Common	Population Unknown	Threats

ID CHECKLIST

- no fin
- slender body
- black with white underside
- white spot under beak
- short, slender beak
- easily startled
- graceful movements
- bouncing swimming motion
- often in mixed schools

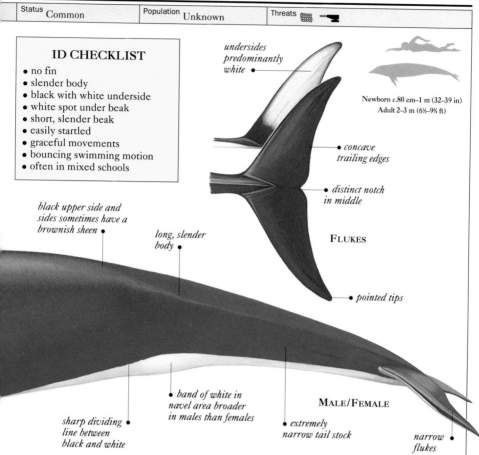

undersides predominantly white •

Newborn *c.*80 cm–1 m (32–39 in)
Adult 2–3 m (6½–9¾ ft)

• *concave trailing edges*

• *distinct notch in middle*

FLUKES

black upper side and sides sometimes have a brownish sheen •

long, slender body •

• *pointed tips*

sharp dividing line between black and white •

• *band of white in navel area broader in males than females*

MALE/FEMALE

• *extremely narrow tail stock*

narrow flukes •

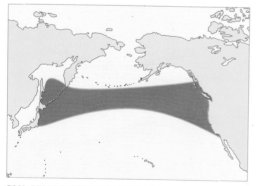

COOL, DEEP TEMPERATE WATERS OF THE NORTHERN NORTH PACIFIC

WHERE TO LOOK

Widely distributed. Occurs in the western North Pacific from Kamchatka, Russia, to Japan, and in the eastern North Pacific from British Columbia, Canada, to northern Baja California, Mexico. May also occur in the northern Sea of Japan. Sometimes ventures farther south when surface temperatures unseasonably low. Seems to be area of very low density immediately south of the Aleutian Islands, Alaska, perhaps separating the eastern and western populations. Mainly found in deep waters over the continental shelf and beyond. May sometimes go inshore, where there are deep canyons. In some areas, may migrate southward and inshore for winter and northward and offshore for summer.

Birth wt Unknown	Adult wt *c.*130–220 lb (60–100 kg)	Diet

Family DELPHINIDAE	Species *Lissodelphis peronii*	Habitat 〰️ (🐋〰️)

SOUTHERN RIGHTWHALE DOLPHIN

The Southern Rightwhale Dolphin is easy to identify at sea. It is the only dolphin in the southern hemisphere without a dorsal fin and also has a striking black and white body pattern. However, at a distance and when swimming at speed, it could be mistaken for a penguin; and when swimming slowly, it could be confused with a fur seal or sea lion. It strongly resembles the Northern Rightwhale Dolphin (p.168), but is slightly smaller and has more white on its head and sides; there is no overlap in range.

Both species were named after the much larger Northern and Southern Right Whales (p.44), which also have no dorsal fins. Calves are grayish brown or cream and develop the adult coloration during their first year. The Southern Rightwhale Dolphin is not particularly well known, mainly because of its predominantly remote, offshore distribution.
• OTHER NAME Mealy-mouthed Porpoise.

slightly flattened body shape may provide stability in absence of dorsal fin •

white forehead in
• *front of blowhole*

clear demarcation between beak and • *forehead*

• *short, white beak*

TEETH $\frac{88-98}{88-98}$

• *eyes located within dark area*

small, curved, predominantly white flippers •

• *pointed tips*

BEHAVIOR
Graceful movements. Often travels very fast in a series of long, low leaps: overall impression is of a bouncing motion rather like a fast-swimming penguin. Sometimes swims slowly, causing little disturbance of the water and exposing only a small part of its head and dark back when surfacing to breathe. Breaching (but with no twisting or turning in the air), belly-flopping, side-slapping, and lobtailing have been observed. Dives may last 6 minutes or more. Some schools will allow close approach, but others flee from boats. Small groups will bow-ride on rare occasions. Often found in the company of Dusky Dolphins, Hourglass Dolphins, or pilot whales. Highly gregarious.

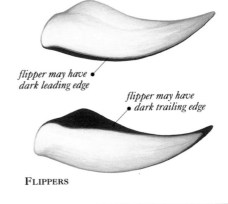

flipper may have • *dark leading edge*

flipper may have • *dark trailing edge*

FLIPPERS

Group size 2–100 (1–1,000)	Fin position No fin

| Status Common | Population Unknown | Threats |

ID CHECKLIST

- no fin
- slender body
- short, white beak
- jet black upper side
- white underside
- clear division between black and white areas
- predominantly white flippers
- bouncing swimming motion

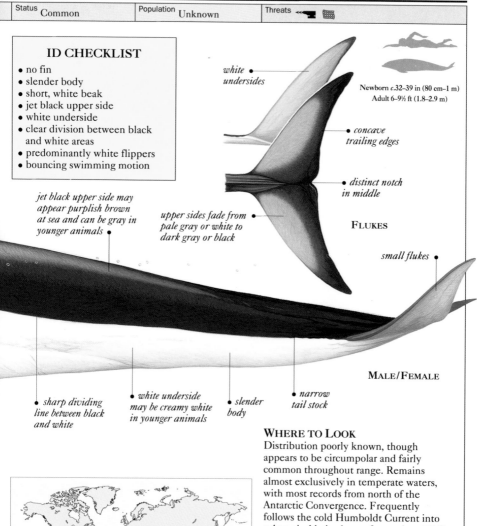

white undersides

Newborn *c.*32–39 in (80 cm–1 m)
Adult 6–9½ ft (1.8–2.9 m)

concave trailing edges

distinct notch in middle

FLUKES

small flukes

jet black upper side may appear purplish brown at sea and can be gray in younger animals

upper sides fade from pale gray or white to dark gray or black

MALE/FEMALE

sharp dividing line between black and white

white underside may be creamy white in younger animals

slender body

narrow tail stock

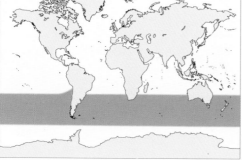

DEEP, COLD TEMPERATE WATERS OF THE SOUTHERN HEMISPHERE

WHERE TO LOOK

Distribution poorly known, though appears to be circumpolar and fairly common throughout range. Remains almost exclusively in temperate waters, with most records from north of the Antarctic Convergence. Frequently follows the cold Humboldt Current into subtropical latitudes, as far north as 19° S off northern Chile, though northernmost record is 12° S off Peru. Southernmost limit varies with sea temperatures from year to year. Seems to be fairly common in the Falklands Current between Patagonia and the Falkland Islands. Believed to occur across the southern Indian Ocean following the West-wind Drift. Seldom seen near land except in sufficiently deep water; however, known to occur in coastal waters off Chile and near New Zealand where water is deeper than 655 ft (200 m).

| Birth wt Unknown | Adult wt *c.*130–220 lb (60–100 kg) | Diet |

Family DELPHINIDAE	Species *Sotalia fluviatilis*	Habitat

TUCUXI

The Tucuxi (pronounced "tookooshee") is one of the smallest of all whales and dolphins. Until recently, there were believed to be up to 5 separate species (*S. brasiliensis, S. fluviatilis, S. guianensis, S. pallida,* and *S. tucuxi*) but these are now considered to be age and color variants of just one, *S. fluviatilis.* Riverine animals are usually paler and smaller than those living along the coast. Many individuals of both forms lighten with age. There may be some confusion with the broadly similar Bottlenose Dolphin (p.192), but the Tucuxi is smaller and has a longer beak and a more triangular dorsal fin with a hooked tip. A large part of its range overlaps with that of the similar-looking Boto (p.226), but again, the Tucuxi is smaller; it also has a more prominent dorsal fin and less pronounced melon. In the southernmost part of its range, it may be impossible to distinguish from a young Franciscana (p.234). Despite its large riverine population, the Tucuxi is not closely related to the "true" river dolphins.

• **OTHER NAME** Estuarine Dolphin.

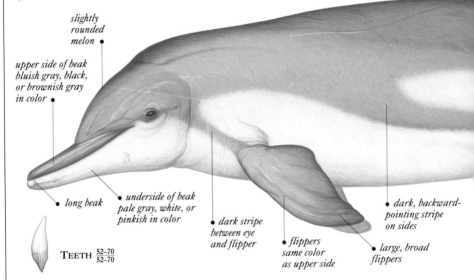

slightly rounded melon

upper side of beak bluish gray, black, or brownish gray in color

long beak

underside of beak pale gray, white, or pinkish in color

TEETH $\frac{52-70}{52-70}$

dark stripe between eye and flipper

flippers same color as upper side

dark, backward-pointing stripe on sides

large, broad flippers

BEHAVIOR

Usually wary of boats, though some individuals may allow a close approach. May surf in waves made by passing boats, but will not bow-ride. Spyhopping, lobtailing, flipper-slapping, and porpoising often seen. Capable of very high breaches (usually falling back on one side), especially after being disturbed. Dives are usually short (around 30 seconds) and it is rarely underwater for longer than a minute. It is an active swimmer. Small groups often swim close together, suggesting strong social ties. It may be seen feeding in the company of river dolphins and, in the Amazon, often associates with feeding terns. Blow is very quiet compared to that of river dolphins. Normally, coastal animals show little of themselves when surfacing, but river inhabitants usually lift the head and part of the body out of water.

ID CHECKLIST
• small size
• robust body
• prominent beak
• slightly rounded melon
• roughly triangular fin
• dark upper side
• light underside
• usually in small groups
• active swimmer

Group size 2–7 (1–30), marine form may occur in largest groups	Fin position Center

Status Locally common	Population Unknown	Threats

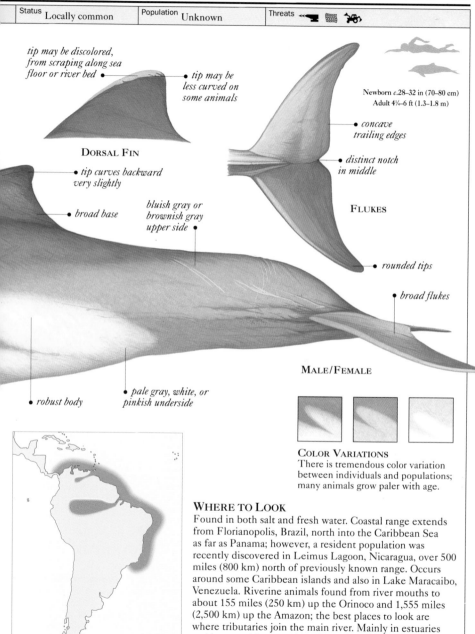

tip may be discolored, from scraping along sea floor or river bed

tip may be less curved on some animals

Newborn *c.*28–32 in (70–80 cm)
Adult 4¼–6 ft (1.3–1.8 m)

concave trailing edges

DORSAL FIN

distinct notch in middle

tip curves backward very slightly

FLUKES

broad base

bluish gray or brownish gray upper side

rounded tips

broad flukes

MALE/FEMALE

robust body

pale gray, white, or pinkish underside

COLOR VARIATIONS
There is tremendous color variation between individuals and populations; many animals grow paler with age.

WHERE TO LOOK
Found in both salt and fresh water. Coastal range extends from Florianopolis, Brazil, north into the Caribbean Sea as far as Panama; however, a resident population was recently discovered in Leimus Lagoon, Nicaragua, over 500 miles (800 km) north of previously known range. Occurs around some Caribbean islands and also in Lake Maracaibo, Venezuela. Riverine animals found from river mouths to about 155 miles (250 km) up the Orinoco and 1,555 miles (2,500 km) up the Amazon; the best places to look are where tributaries join the main river. Mainly in estuaries and bays, and in deep river channels or floodplain lakes.

SHALLOW COASTAL WATERS AND RIVERS OF NORTHEASTERN SOUTH AMERICA AND EASTERN CENTRAL AMERICA

Birth wt Unknown	Adult wt 75–100 lb (35–45 kg)	Diet

Family DELPHINIDAE	Species *Sousa chinensis*	Habitat 〰️

INDO-PACIFIC HUMP-BACKED DOLPHIN

Classification of the hump-backed dolphins is still in dispute. There could be as many as 5 species, although most authorities currently accept only 2: the Indo-Pacific and the Atlantic. However, there certainly appear to be 2 distinct populations of the Indo-Pacific Hump-backed Dolphin: one type is found west of Sumatra, Indonesia; the other to the east and south of the island. Animals in the west have a characteristic fatty hump, while those in the east have a more prominent dorsal fin but no hump. Confusion is possible with the Bottlenose Dolphin (p.192), but the unusual surfacing motion of the Indo-Pacific Hump-backed should be distinctive east of Sumatra, and its humped back is usually sufficient for identification in the west.

• **OTHER NAMES** Indo-Pacific Humpback Dolphin, Speckled Dolphin, *S. plumbea* (animals west of Sumatra).

elongated hump in middle of back (variable) •

slightly rounded • *melon*

tip of beak may lighten • *with age*

long, slender beak •

fairly straight mouth line •

body may • *be speckled*

TEETH $\frac{58-76}{58-76}$

broad flippers with rounded tips

• *underside usually lighter than sides and upper side*

BEHAVIOR

Usually quite difficult to approach and tends to avoid boats by diving and reappearing some distance away in a different direction. Rarely bow-rides. Distinctive surfacing behavior: breaks surface at an angle of 30° to 45°, clearly showing beak and sometimes entire head, and a few seconds later arches back strongly and may lift flukes into the air. Surfaces about every 40 to 60 seconds but can stay underwater for several minutes. Normally a slow swimmer, but courtship may involve chasing one another around in circles at high speed. May turn on one side and wave a flipper in the air. Sometimes spyhops. Often breaches, especially when young, and may do complete back somersaults. May lobtail when feeding. Associates with Bottlenose Dolphins and, to a lesser extent, with Finless Porpoises and Long-snouted Spinner Dolphins.

ID CHECKLIST

• robust body
• elongated hump on back (animals west of Sumatra only)
• small fin sits on hump
• long, slender beak
• beak exposed on surfacing
• back strongly arched on diving
• flukes raised on diving
• difficult to approach

Group size 3–7 (1–25), small groups may congregate to form larger groups	Fin position Center

Status Locally common	Population Unknown	Threats

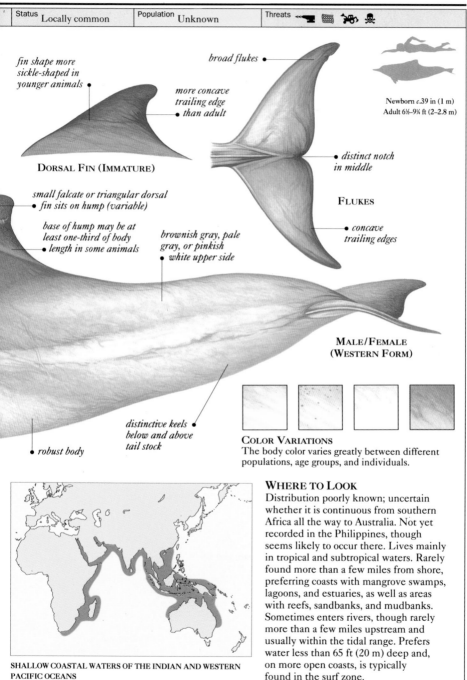

fin shape more
sickle-shaped in
younger animals •

broad flukes •

Newborn *c*.39 in (1 m)
Adult 6½–9¾ ft (2–2.8 m)

• *more concave
trailing edge
• than adult*

DORSAL FIN (IMMATURE)

• *distinct notch
in middle*

FLUKES

*small falcate or triangular dorsal
• fin sits on hump (variable)*

*base of hump may be at
least one-third of body
• length in some animals*

*brownish gray, pale
gray, or pinkish
• white upper side*

• *concave
trailing edges*

**MALE/FEMALE
(WESTERN FORM)**

*distinctive keels •
below and above
tail stock*

• *robust body*

COLOR VARIATIONS
The body color varies greatly between different
populations, age groups, and individuals.

WHERE TO LOOK
Distribution poorly known; uncertain
whether it is continuous from southern
Africa all the way to Australia. Not yet
recorded in the Philippines, though
seems likely to occur there. Lives mainly
in tropical and subtropical waters. Rarely
found more than a few miles from shore,
preferring coasts with mangrove swamps,
lagoons, and estuaries, as well as areas
with reefs, sandbanks, and mudbanks.
Sometimes enters rivers, though rarely
more than a few miles upstream and
usually within the tidal range. Prefers
water less than 65 ft (20 m) deep and,
on more open coasts, is typically
found in the surf zone.

SHALLOW COASTAL WATERS OF THE INDIAN AND WESTERN
PACIFIC OCEANS

Birth wt *c*.55 lb (25 kg)	Adult wt 330–440 lb (150–200 kg)	Diet

Family DELPHINIDAE	Species *Sousa teuszii*	Habitat 〰️

ATLANTIC HUMP-BACKED DOLPHIN

Some authorities believe the Atlantic Hump-backed Dolphin to be a geographical variant of the Indo-Pacific Hump-backed Dolphin (p.174). However, in the light of current evidence, many consider this unlikely as there are morphological differences between the two (mainly in the number of teeth and vertebrae). Confusion is most likely with the Bottlenose Dolphin (p.192), but with a conspicuous, elongated hump in the middle of its back, and a relatively small dorsal fin on top, the Atlantic Hump-backed Dolphin is fairly easy to recognize. It is well known for cooperating with Mauritanian fishermen, around Cap Timiris, north of Nouakchott, by driving fish toward their nets.

• **OTHER NAMES** Atlantic Humpback Dolphin, Cameroon Dolphin.

base of hump may be at least one-third of body length

conspicuous, elongated hump on back, on adults only

slightly rounded melon

long, slender beak

tip of beak may lighten with age

fairly straight mouth line

body may be speckled

broad flippers with rounded tips

underside usually paler than sides and upper side

TEETH $\frac{52-62}{52-62}$

BEHAVIOR
Usually difficult to approach and tends to avoid boats by diving and reappearing some distance away in a different direction. Rarely bow-rides. Surfaces about every 40 to 60 seconds but can stay underwater for several minutes. Unusual surfacing behavior like that of Indo-Pacific Hump-backed Dolphin. Normally a slow swimmer, but courtship may involve chasing one another around in circles at high speed. May turn on one side and wave a flipper in the air. Sometimes spyhops. Frequently breaches (especially young animals) and may do complete back somersaults. May associate with Bottlenose Dolphins.

ID CHECKLIST
• robust body
• elongated hump on back
• small fin sits on hump
• long, slender beak
• beak exposed on surfacing
• back strongly arched on diving
• flukes raised on diving
• usually in small groups
• difficult to approach

Group size 3–7 (1–25), small groups may congregate to form larger groups	Fin position Center

Status Locally common	Population Unknown	Threats

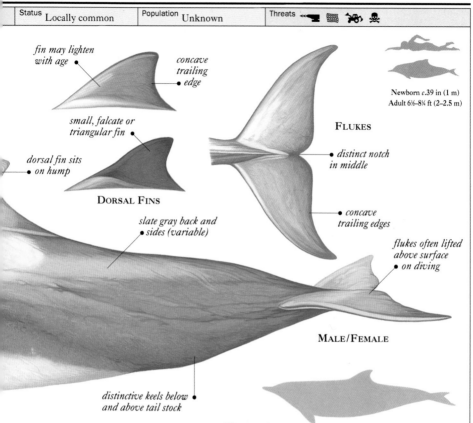

fin may lighten with age

concave trailing edge

Newborn *c*.39 in (1 m)
Adult 6½–8¼ ft (2–2.5 m)

FLUKES

small, falcate or triangular fin

dorsal fin sits on hump

distinct notch in middle

DORSAL FINS

slate gray back and sides (variable)

concave trailing edges

flukes often lifted above surface on diving

MALE/FEMALE

distinctive keels below and above tail stock

YOUNG ANIMALS
Young Atlantic Hump-backed Dolphins have a less pronounced melon and a more sickle-shaped dorsal fin than adults, and no hump in the middle of the back. They tend to darken in color as they grow older.

WHERE TO LOOK
Distribution based on scant evidence and may be more extensive than few records suggest. Known range extends along coast of West Africa, from Mauritania to Cameroon and possibly as far south as Angola. Appears to be isolated from the very similar Indo-Pacific Hump-backed Dolphin (p.174) living along the South African coast. Seems to be particularly common in southern Senegal and north-western Mauritania. Prefers shallow coastal and estuarine water less than 65 ft (20 m) deep, especially around mangrove swamps. Typically occurs in the surf zone on more open coasts. Known to enter Niger and Bandiala rivers, and possibly others, though rarely travels far upstream and usually remains within the tidal range.

COASTAL WATERS OF TROPICAL WEST AFRICA

Birth wt Unknown	Adult wt 220–330 lb (100–150 kg)	Diet

Family DELPHINIDAE	Species *Stenella coeruleoalba*	Habitat 〰️ (🏝️)

STRIPED DOLPHIN

The Striped Dolphin is fairly easy to identify at sea by its distinctive stripe; some individuals also have bright pink undersides. At first glance, it may resemble a Common Dolphin (p.164), which is broadly similar in size and shape; however, the Striped Dolphin has a dark body stripe and, unlike the Common Dolphin, it does not have a yellow hourglass pattern on its sides. Confusion is also possible with Fraser's Dolphin (p.208), although the Striped Dolphin is more streamlined and has a longer beak and a larger, more falcate dorsal fin. One of its most distinctive features is the pale gray finger-shaped marking below the dorsal fin, but this is also characteristic of many Atlantic Spotted and Bottlenose Dolphins. It is a common species, but its population has declined in recent years.

• **OTHER NAMES** Euphrosyne Dolphin, Whitebelly, Blue-white Dolphin, Meyen's Dolphin, Gray's Dolphin, Streaker Porpoise.

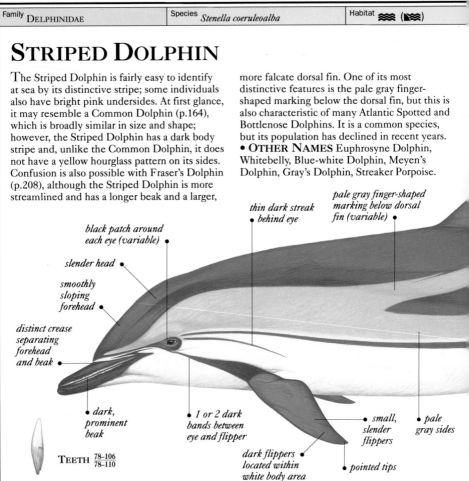

pale gray finger-shaped marking below dorsal fin (variable) •

thin dark streak • behind eye

black patch around each eye (variable) •

slender head •

smoothly sloping forehead •

distinct crease separating forehead and beak •

• dark, prominent beak

TEETH 78–106 / 78–110

• 1 or 2 dark bands between eye and flipper

dark flippers • located within white body area

• small, slender flippers

• pale gray sides

• pointed tips

BEHAVIOR

Active and highly conspicuous. Frequently breaches, sometimes as high as 23 ft (7 m); capable of amazing acrobatics, including back somersaults, tail-spins, and upside down porpoising. When swimming at speed, up to one-third of all members of a school will be above the surface at any one time. Dives typically last 5 to 10 minutes. When feeding, dives to at least 655 ft (200 m) deep. Will bow-ride in some areas (mainly the Atlantic and Mediterranean) but rarely approaches vessels in other areas. Smaller groups (under 100) tend to occur in the Atlantic and Mediterranean. Often associates with Common Dolphins and, in the eastern tropical Pacific, with Yellowfin Tuna. Several mass strandings have occurred in recent years.

ID CHECKLIST

- dark, prominent fin
- long, dark side stripe
- dark stripe from eye to flipper
- pale "finger" marking below fin
- prominent beak
- slender body
- white or pink underside
- usually in large schools
- active at surface

Group size 10–500 (1–3,000)	Fin position Center

| Status Common | Population Unknown | Threats  |

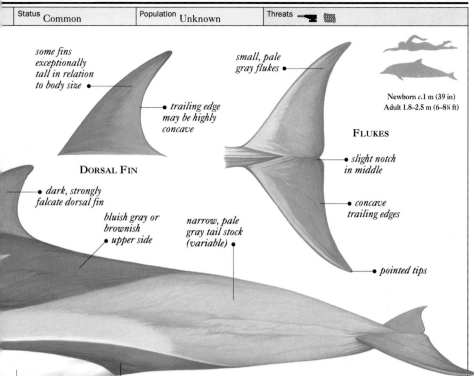

*some fins
exceptionally
tall in relation
to body size* •

*small, pale
gray flukes* •

Newborn *c.*1 m (39 in)
Adult 1.8–2.5 m (6–8¼ ft)

FLUKES

• *trailing edge
may be highly
concave*

DORSAL FIN

• *slight notch
in middle*

• *dark, strongly
falcate dorsal fin*

*bluish gray or
brownish
• upper side*

*narrow, pale
gray tail stock
(variable)* •

• *concave
trailing edges*

• *pointed tips*

MALE/FEMALE

• *white or pink
underside*

*thin, dark stripe running •
from underside of tail stock
to eye (variable)*

COLOR VARIATIONS
The main body color and striping are highly
variable; the upper side, for example, can be
anything from bluish gray to brownish gray.

WHERE TO LOOK
Mainly tropical and subtropical, though
also found in warm temperate waters.
Wide distribution, though it does not
appear to be continuous: there are
gaps and low densities in some areas,
suggesting several geographically isolated
(or semi-isolated) populations. Distinctive
seasonal migration recorded off the coast
of Japan, where it has been well studied:
appears to winter in the East China Sea
and summer in the pelagic North Pacific.
Migrations are unknown in other parts of
the world, though it may move seasonally
with warm oceanic currents in some areas.
Primarily occurs offshore and, when found
close to land, usually in deep water.

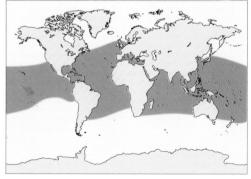

WARM TEMPERATE, SUBTROPICAL, AND TROPICAL WATERS
AROUND THE WORLD

| Birth wt Unknown | Adult wt 200–330 lb (90–150 kg) | Diet |

Family DELPHINIDAE	Species *Stenella clymene*	Habitat 〜〜

SHORT-SNOUTED SPINNER DOLPHIN

For many years, the Short-snouted Spinner Dolphin was considered to be one of many variations of the Long-snouted Spinner Dolphin (p.182), but it was officially classified as a separate species in 1981. There is considerable overlap in range between the two species in the Atlantic, and they can be difficult to tell apart at sea. The Short-snouted Spinner is slightly more robust than the Long-snouted Spinner, its dorsal fin is less triangular, and as the name suggests, it has a shorter, slightly stubbier beak; also, look for the dark dorsal cape of the Short-snouted Spinner, which dips down below the

dorsal fin and nearly touches the white underside. There may also be confusion with Bottlenose Dolphins (p.192) and Common Dolphins (p.164). The Short-snouted Spinner Dolphin is probably not very abundant, though it may have been overlooked in some areas because of identification difficulties.
• **OTHER NAMES** Clymene Dolphin, Helmet Dolphin, Senegal Dolphin.

dark gray or black cape dips below dorsal fin •

pale gray stripe between blowhole and beak •

slightly bulging forehead •

• pointed tips

• lower jaw white (variable)

pale gray • stripe from eye to flipper (variable)

• dark, slender flippers (variable)

TEETH $\frac{78-98}{76-96}$

• tip of beak and "lips" black

face markings variable, but may be similar to Short-snouted Spinner

longer, more slender beak than Short-snouted Spinner •

BEHAVIOR
Sometimes spins longitudinally when breaching, normally landing on back or one side; recent observations in the Gulf of Mexico indicate that the leaps of the Short-snouted Spinner are as high and complex as in the Long-snouted Spinner, but this appears to be rare in most populations. Short-snouted Spinner known to bow-ride in some areas and sometimes approaches boats quite closely. Thought to be a midwater, nocturnal feeder. May be seen in association with Long-snouted Spinners and Common Dolphins and various small whales.

LONG-SNOUTED SPINNER DOLPHIN

Group size 5–50 (1–500)	Fin position Center

Status Unknown	Population Unknown	Threats Unknown

ID CHECKLIST

- slightly falcate fin
- black-tipped beak
- black "lips"
- dark gray or black cape
- cape dips below fin
- 3-toned color pattern
- fairly robust body
- prominent beak
- dark, slender flippers

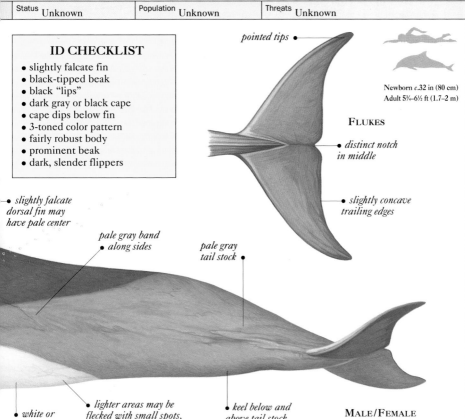

pointed tips

Newborn *c*.32 in (80 cm)
Adult 5¾–6½ ft (1.7–2 m)

FLUKES

• *distinct notch in middle*

• *slightly concave trailing edges*

— • *slightly falcate dorsal fin may have pale center*

• *pale gray band along sides*

• *pale gray tail stock*

• *slightly falcate dorsal fin may have pale center*

• *white or pinkish underside*

• *lighter areas may be flecked with small spots, especially where white and gray shading meet*

• *keel below and above tail stock*

MALE/FEMALE

WHERE TO LOOK

Distribution poorly known, and this map is based on a relatively small number of sightings. Mostly found in tropical and subtropical waters, and occasionally in warm temperate waters. Recorded off the northwest coast of Africa; in the mid-Atlantic around the equator; along the northeastern coast of South America; in southeastern USA as far north as New Jersey (which is the northernmost record for the species); in the Gulf of Mexico; and in the Caribbean Sea. May occur as far south as southern Brazil in the west of the range (although single record in 1992 in state of Santa Catarina, Brazil, could have been a stray) and Angola in the east, but limits of distribution are not known for certain. Found mainly in deep water.

TROPICAL, SUBTROPICAL, AND OCCASIONALLY WARM TEMPERATE WATERS OF THE ATLANTIC OCEAN

Birth wt Unknown	Adult wt 110–200 lb (50–90 kg)	Diet

Family DELPHINIDAE	Species *Stenella longirostris*	Habitat ≈≈≈ ▶≈≈

LONG-SNOUTED SPINNER DOLPHIN

One of the most acrobatic of all cetaceans, the Long-snouted Spinner Dolphin is well known for its spectacular aerial displays. There are many varieties, differing greatly in body shape, size, and color. Four live in the eastern tropical Pacific alone (the Hawaiian, eastern, Costa Rican, and whitebelly forms) and there are other, lesser known, varieties elsewhere in the world. The most recent discovery is a "dwarf" form found in the Gulf of Thailand. The best way to distinguish all Long-snouted Spinner

Dolphins from other species is by the long, slender beak, the erect dorsal fin, and their high, spinning leaps; most also have a distinctive 3-toned color pattern, though eastern Pacific animals are mainly gray. Hundreds of thousands of Long-snouted Spinners have been killed by tuna purse-seine fisheries in the eastern tropical Pacific, causing a great decline in the population of this area in recent years.

• **OTHER NAMES** Longsnout, Spinner, Long-beaked Dolphin, Rollover.

crease where beak joins forehead •

distinct but gently • sloping forehead

predominantly dark gray body •

dorsal fin leans forward in large males of eastern tropical Pacific •

long, thin beak •

• dark-tipped beak

• black "lips"

dark gray stripe • from eye to flipper

TEETH $\frac{88-128}{84-124}$

long, pointed flippers •

long, pointed flippers •

BEHAVIOR

When breaching, hurls itself up to 9¾ ft (3 m) into the air, then twists its body into sinuous curves or spins round on its longitudinal axis up to 7 times in a single leap. Short-snouted Spinner Dolphin (p.180) is the only other cetacean that does this (others somersault, but do not spin on the longitudinal axis). May also be seen breaching in the normal way. Readily bow-rides in most areas (will come to bow of a boat from far away and may stay for half an hour or more) but much more nervous in eastern tropical Pacific and rarely approaches boats in Lesser Antilles, Caribbean. Large schools often churn water into a foam when swimming. Often associates with Pantropical Spotted Dolphins, Yellowfin Tuna, and seabirds in the eastern tropical Pacific; may be seen with other cetaceans throughout range.

triangular or slightly falcate fin •

3 distinct patches of • color

WHITEBELLY FORM

slightly falcate fin •

3 distinct areas • of coloration

HAWAIIAN FORM

• slightly larger and darker than whitebelly form

Group size 5–200 (1–1,000), sometimes in larger, mixed schools	Fin position Center

Status Common	Population Unknown	Threats

ID CHECKLIST

- performs high, spinning leaps
- slender body
- long, slender beak
- long, pointed flippers
- tall, erect fin
- dark-tipped beak
- 3-toned color pattern
- gently sloping forehead
- usually lives in large schools

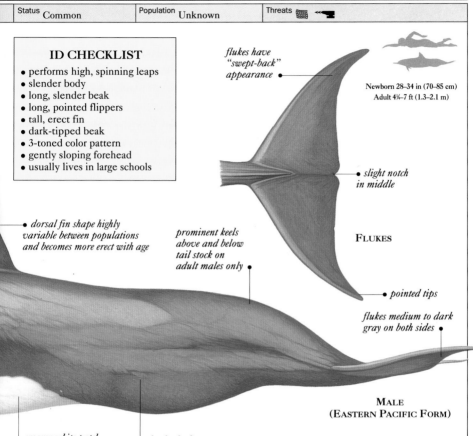

flukes have "swept-back" appearance •

Newborn 28–34 in (70–85 cm)
Adult 4¼–7 ft (1.3–2.1 m)

• slight notch in middle

FLUKES

• dorsal fin shape highly variable between populations and becomes more erect with age

prominent keels above and below tail stock on adult males only •

• pointed tips

flukes medium to dark gray on both sides •

MALE (EASTERN PACIFIC FORM)

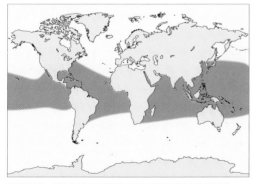

• creamy white patch on belly (variable)

• slender body

WHERE TO LOOK

Sometimes occurs in warm temperate waters, but mainly tropical. Each variety has a more limited range than the species as a whole: for example, the Costa Rican form is found only in a narrow band of water less than 95 miles (150 km) wide off western Central America; and the eastern form is found from the tip of Baja California, Mexico, south to the equator and offshore to about 125°. Two or more varieties may occur in the same area. Distribution in the Atlantic poorly known. Most common far out to sea, especially in the eastern tropical Pacific, but also found close to shore, for example off southeastern USA and around some islands. The Hawaiian form seems to rest inshore by day and moves to feed offshore at night.

TROPICAL AND SUBTROPICAL WATERS IN THE ATLANTIC, INDIAN, AND PACIFIC OCEANS

Birth wt Unknown	Adult wt 100–165 lb (45–75 kg)	Diet

Family DELPHINIDAE	Species *Stenella attenuata*	Habitad ≈≈≈ ▰≋≋

PANTROPICAL SPOTTED DOLPHIN

The Pantropical Spotted Dolphin can vary greatly in size, shape, and color. Two main forms are recognized: one living along the coast, the other offshore. The coastal form is generally the larger, more robust, and more spotted of the two. Most adult animals can be identified by their spotting, although the spots are virtually absent in certain populations, such as around Hawaii and in the Gulf of Mexico. Confusion may occur with the Bottlenose Dolphin (p.192) and hump-backed dolphins (pp.174–177), which may also have some spotting, and it can be very difficult to distinguish from the Atlantic Spotted Dolphin (p.186) in parts of the Atlantic Ocean. It is probably one of the most common cetaceans although, since the early 1960s, incidental catches in Yellowfin Tuna nets have reduced some populations in the eastern tropical Pacific by as much as 65 percent.
• **OTHER NAMES** Spotted Dolphin, White-spotted Dolphin, Bridled Dolphin, Spotter, Spotted Porpoise, Slender-beaked Dolphin.

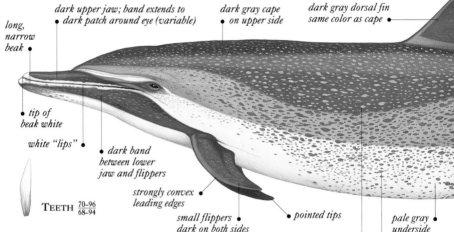

long, narrow beak •

dark upper jaw; band extends to
• *dark patch around eye (variable)*

dark gray cape
• *on upper side*

dark gray dorsal fin
same color as cape •

• *tip of beak white*

white "lips" •
• *dark band between lower jaw and flippers*

TEETH $\frac{70-96}{68-94}$

strongly convex leading edges •

small flippers •
dark on both sides

• *pointed tips*

light spots cover •
dark areas of body

• *dark spots cover light areas of body*

pale gray •
underside

BEHAVIOR
Very active: schools may be sighted from afar by froth caused by their leaping. It is a fast, energetic swimmer, using long, shallow leaps. Frequently breaches, sometimes hurling itself high into the air, where it seems to hang before falling back with a splash. Often associates with Long-snouted Spinner Dolphins and Yellowfin Tuna, and often seen with feeding seabirds. Lobtailing and bow-riding common, but in tuna fishing areas some individuals flee from boats.

COLOR VARIATIONS
The amount of spotting varies according to age and location. Newborn animals are unspotted. Juveniles develop dark spots on the underside, followed by light spots on the upper side; these increase in number and size with age. Some older animals are so spotted that the background color is barely visible, and their upper sides may be so pale that they are nick-named "silverbacks."

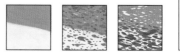

ID CHECKLIST
• dark gray cape
• dark line from flipper to beak
• tall, falcate fin
• slender, elongated body
• long, narrow beak
• white-tipped beak and "lips"
• most adults heavily spotted
• appearance varies within school
• very active at surface

Group size 50–1,000 (5–3,000), coastal form usually in groups of under 100	Fin position Center

Status Common	Population Unknown	Threats

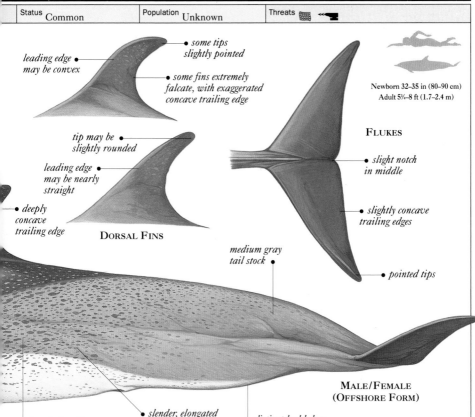

Newborn 32–35 in (80–90 cm)
Adult 5¾–8 ft (1.7–2.4 m)

some tips slightly pointed

leading edge may be convex

some fins extremely falcate, with exaggerated concave trailing edge

tip may be slightly rounded

leading edge may be nearly straight

deeply concave trailing edge

DORSAL FINS

FLUKES

slight notch in middle

slightly concave trailing edges

pointed tips

medium gray tail stock

MALE/FEMALE (OFFSHORE FORM)

band of medium gray along each side (may be absent)

slender, elongated body (coastal form more robust)

distinct keel below (and sometimes above) tail stock, except on large males

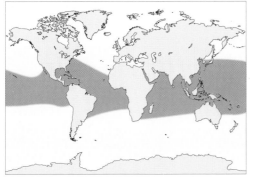

TROPICAL AND SOME WARM TEMPERATE WATERS OF THE
ATLANTIC, PACIFIC, AND INDIAN OCEANS

WHERE TO LOOK

Widely distributed, mainly in tropical seas but also in the subtropics and some warm temperate waters. Distribution probably not continuous within the range, though it appears to be abundant in many areas. Found mainly where surface water temperature higher than 77° F (25° C). Commonly occurs around islands. Well-studied in the eastern tropical Pacific but poorly known elsewhere. Overlaps with the Atlantic Spotted Dolphin, mainly in the western North Atlantic, where it occurs mainly offshore. No known migrations, though offshore form may make seasonal movements, usually summering inshore and wintering offshore.

Birth wt Unknown	Adult wt 200–255 lb (90–115 kg)	Diet

Family DELPHINIDAE	Species *Stenella frontalis*	Habitat 〰〰 〰

ATLANTIC SPOTTED DOLPHIN

The Atlantic Spotted Dolphin has been studied widely in the western North Atlantic, but is poorly known in other areas. It closely resembles the Pantropical Spotted Dolphin (p.184), but has a slightly more robust body and a light streak on each shoulder, as well as spots on the underside that remain distinctly defined and rarely merge. Confusion is also possible with Atlantic Hump-backed Dolphins (p.176), Rough-toothed Dolphins (p.190), Bottlenose Dolphins (p.192), and spinner dolphins (pp.180–183). In schools of mixed age groups, the Atlantic Spotted Dolphin can be identified by the extensive spotting in older animals (although the spotting is variable and can be hard to see in certain lights) and the dark cape of younger animals (which have no spots). There are so many variations of spotted dolphin that its taxonomy has puzzled experts for a long time; however, the Atlantic Spotted Dolphin is now accepted as a separate species.

• **OTHER NAMES** Spotted Porpoise, Spotter, Bridled Dolphin, Gulf Stream Spotted Dolphin, Long-snouted Dolphin; formerly *S. plagiodon* (in eastern USA).

dark purplish gray cape on upper side

tall, falcate dorsal fin (variable)

fairly robust head and body

"lips" may be white

moderately long, chunky beak tipped with white

faint light gray band between eye and flipper

TEETH $\frac{64-84}{60-80}$

curved flippers usually unspotted

pale diagonal shoulder blaze (variable)

pointed tips

BEHAVIOR
Very active at the surface. Often breaches, sometimes hurling itself high into the air, where it seems to hang before falling back with a splash; most aerial behavior observed when feeding. Fast and energetic swimmer, using long, shallow leaps. Avid bow-rider: may swim far to join a fast-moving vessel (though more wary where hunted). Frequent reports of mixed schools with Bottlenose Dolphins may be cases of mistaken identity (possibly a result of older animals being spotted and younger ones unspotted). When surfacing, tip of the beak typically breaks the surface first, followed by the head, back, and dorsal fin. Group size is generally smaller (5–15) in inshore populations. Social structure appears to be fairly complex and is believed to include individual recognition and bonding.

ID CHECKLIST
- fairly robust head and body
- most adults heavily spotted
- pale diagonal shoulder blaze
- long, chunky, white-tipped beak
- 3-toned color pattern
- tall, falcate fin
- dark purplish gray cape
- variable appearance within herd
- very active at surface

Group size 5–15 (1–50), a few hundred may form temporary gatherings	Fin position Center

Status Locally common	Population Unknown	Threats 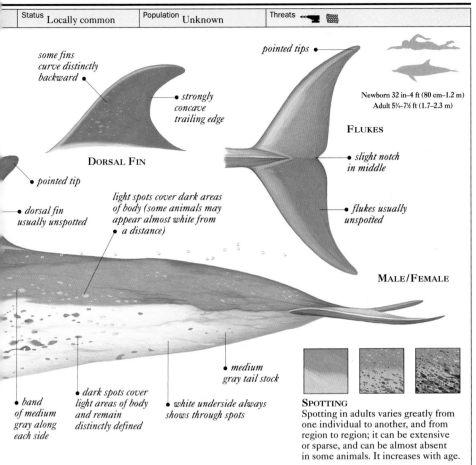

some fins curve distinctly backward

pointed tips

strongly concave trailing edge

Newborn 32 in–4 ft (80 cm–1.2 m)
Adult 5¾–7½ ft (1.7–2.3 m)

FLUKES

DORSAL FIN

pointed tip

slight notch in middle

dorsal fin usually unspotted

light spots cover dark areas of body (some animals may appear almost white from a distance)

flukes usually unspotted

MALE/FEMALE

medium gray tail stock

band of medium gray along each side

dark spots cover light areas of body and remain distinctly defined

white underside always shows through spots

SPOTTING
Spotting in adults varies greatly from one individual to another, and from region to region; it can be extensive or sparse, and can be almost absent in some animals. It increases with age.

WHERE TO LOOK
Known only from the Atlantic, where it occurs mainly in warm waters. Distribution off South America and West Africa poorly known and may be more extensive than map shows. Appears to be common in the western North Atlantic and the Gulf of Mexico. In the eastern North Atlantic, occurs further north than the map suggests, with many recent records from around the Azores and possible sightings around the Canary Islands. Gulf of Mexico population (and possibly other populations) moves closer to shore during summer. Usually found over the offshore continental shelf. Smaller, less spotted form more pelagic than larger, heavily spotted form.

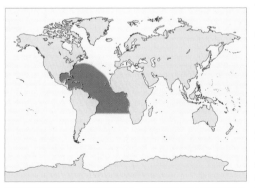

WARM TEMPERATE, SUBTROPICAL, AND TROPICAL WATERS IN BOTH THE NORTH AND SOUTH ATLANTIC

Birth wt Unknown	Adult wt 220–310 lb (100–140 kg)	Diet

Family DELPHINIDAE	Species *Stenella frontalis*	Habitat 〰 🐟

VARIATIONS

Atlantic Spotted Dolphins show tremendous variation in color and extent of spotting, and no two look exactly alike. Two main forms are recognized; one lives inshore and the other offshore. The inshore form is generally larger, more robust, and more heavily spotted than the offshore form (spotting usually decreases with distance from the mainland, and from west to east across the Atlantic); inshore animals also have wider beaks and larger teeth (possibly because they normally take larger prey). In both forms, the spots become bigger and more numerous with age and are most extensive in large, old animals. A good way to distinguish the Atlantic Spotted Dolphin from the Pantropical Spotted Dolphin is by the extent of dark spotting on the underside; the spots remain distinctly defined in the Atlantic Spotted but merge to hide the background color in the Pantropical Spotted. Spotting is not unique to these two species: some Bottlenose Dolphins have a moderate number of spots; Rough-toothed Dolphins often have pinkish white or yellowish white blotches; the paler areas of Short-snouted Spinner Dolphins are often flecked with small spots; and hump-backed dolphins are sometimes speckled. Some species also have white scarring on their bodies, which can resemble spotting or flecking.

OLD ANIMAL

• *background color may be partly obscured by extensive spotting*

NEWBORN

The Atlantic Spotted Dolphin is born with a distinctive dark gray or purplish gray cape, light gray sides, a white underside, and no spots. When it is about 1 year old, grayish white spots first begin to appear, normally low down on the sides; these spots spread upward toward the cape (which becomes less distinct with increasing age), while dark spotting begins to develop on the underside. By the end of the animal's second year, spots cover most of the body and continue to increase in number as it grows older.

flukes have dark trailing edges •

• *dark cape distinctly separate from lighter sides*

• *no spots*

• *flippers darker than rest of body*

Group size 5–15 (1–50), a few hundred may form temporary gatherings	Fin position Center

Status Locally common	Population Unknown	Threats 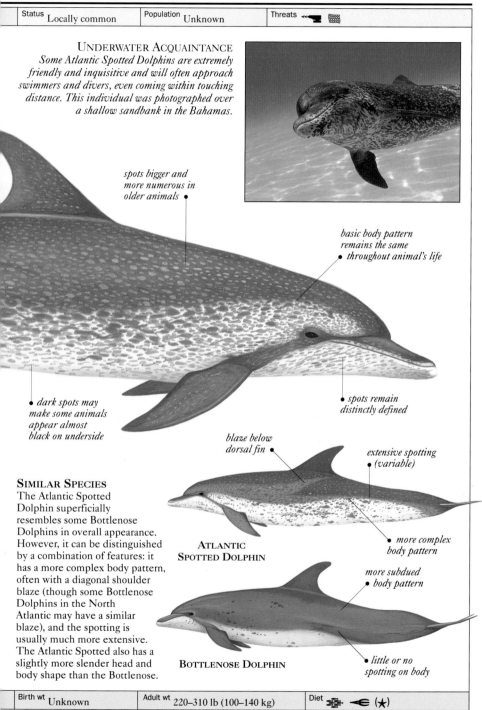

UNDERWATER ACQUAINTANCE
Some Atlantic Spotted Dolphins are extremely friendly and inquisitive and will often approach swimmers and divers, even coming within touching distance. This individual was photographed over a shallow sandbank in the Bahamas.

spots bigger and more numerous in older animals

basic body pattern remains the same throughout animal's life

dark spots may make some animals appear almost black on underside

spots remain distinctly defined

blaze below dorsal fin

extensive spotting (variable)

SIMILAR SPECIES
The Atlantic Spotted Dolphin superficially resembles some Bottlenose Dolphins in overall appearance. However, it can be distinguished by a combination of features: it has a more complex body pattern, often with a diagonal shoulder blaze (though some Bottlenose Dolphins in the North Atlantic may have a similar blaze), and the spotting is usually much more extensive. The Atlantic Spotted also has a slightly more slender head and body shape than the Bottlenose.

ATLANTIC SPOTTED DOLPHIN

more complex body pattern

more subdued body pattern

BOTTLENOSE DOLPHIN

little or no spotting on body

Birth wt Unknown	Adult wt 220–310 lb (100–140 kg)	Diet

Family DELPHINIDAE	Species *Steno bredanensis*	Habitat 〰〰

ROUGH-TOOTHED DOLPHIN

The Rough-toothed Dolphin is relatively easy to identify at sea, but it is rarely seen and is poorly known. Its head has a unique shape: the long, narrow beak blends into the forehead without a crease, unlike all other dolphins with prominent beaks. Also, the narrowness of the head and the unusually large eyes give it a slightly reptilian appearance. Although it is a very distinctive animal, confusion is still possible with several other species, especially Bottlenose Dolphins (p.192), spotted dolphins (pp.184–189), and spinner dolphins

(pp.180–183). In addition to the head shape, look out for its dark cape, white "lips," and yellowish white or pinkish white blotches. Dead animals can be identified by the fine vertical wrinkles on the teeth (hence the common name), but these are often difficult to detect. There may be some variation in shape between populations, most notably between Atlantic and Indo-Pacific animals.

• **OTHER NAME** Slopehead.

dark gray or bluish gray cape may have purplish hue

body robust in front of dorsal fin

smoothly sloping forehead without crease between beak and forehead

conical head

long, narrow beak

white or pinkish white "lips" and throat (variable)

dark patch around large eyes

TEETH $\frac{38-52}{38-56}$

sides paler than upper side and may have purplish hue

large, pointed flippers

white or pinkish white underside

BEHAVIOR

Difficult to observe as may stay submerged for as long as 15 minutes. Rarely does more than a half-hearted breach. Fast swimmer, sometimes porpoising with low, arc-shaped leaps. May swim rapidly just under the surface, with dorsal fin and small part of back clearly visible. Sometimes bow-rides, especially in front of fast-moving vessels, though not as readily as many other tropical dolphins. May associate with Bottlenose Dolphins and pilot whales and, less frequently, with spinner dolphins and spotted dolphins; sometimes associates with shoals of Yellowfin Tuna. May be seen logging.

ID CHECKLIST

- tall, falcate fin
- conical head
- beak continuous with forehead
- dark, narrow cape
- pinkish white blotches
- long, narrow beak
- white "lips"
- white or pinkish white below
- usually in small groups

Group size 10–20 (1–50), occasionally in groups of several hundred	Fin position Center

Status Unknown	Population Unknown	Threats

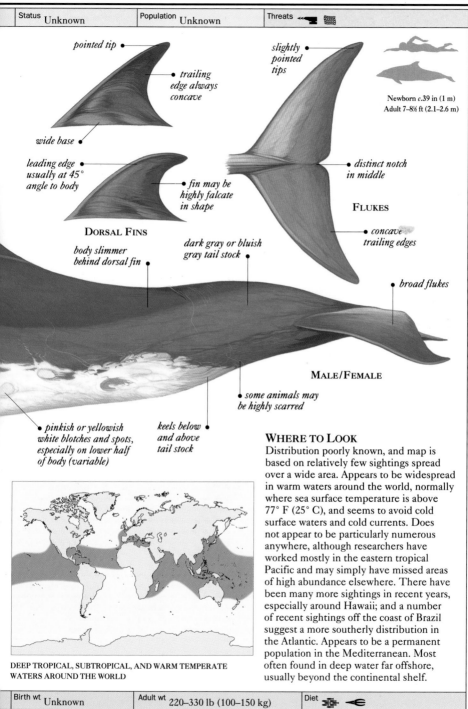

pointed tip •

• trailing
edge always
concave

slightly •
pointed
tips

Newborn *c.*39 in (1 m)
Adult 7–8½ ft (2.1–2.6 m)

• wide base

leading edge •
usually at 45°
angle to body

• fin may be
highly falcate
in shape

DORSAL FINS

body slimmer
behind dorsal fin •

dark gray or bluish
gray tail stock •

• distinct notch
in middle

FLUKES

• concave
trailing edges

• broad flukes

MALE/FEMALE

• some animals may
be highly scarred

• pinkish or yellowish
white blotches and spots,
especially on lower half
of body (variable)

keels below •
and above
tail stock

WHERE TO LOOK

Distribution poorly known, and map is based on relatively few sightings spread over a wide area. Appears to be widespread in warm waters around the world, normally where sea surface temperature is above 77° F (25° C), and seems to avoid cold surface waters and cold currents. Does not appear to be particularly numerous anywhere, although researchers have worked mostly in the eastern tropical Pacific and may simply have missed areas of high abundance elsewhere. There have been many more sightings in recent years, especially around Hawaii; and a number of recent sightings off the coast of Brazil suggest a more southerly distribution in the Atlantic. Appears to be a permanent population in the Mediterranean. Most often found in deep water far offshore, usually beyond the continental shelf.

DEEP TROPICAL, SUBTROPICAL, AND WARM TEMPERATE
WATERS AROUND THE WORLD

Birth wt Unknown	Adult wt 220–330 lb (100–150 kg)	Diet

| Family DELPHINIDAE | Species *Tursiops truncatus* | Habitat 〰️ 〰️ |

BOTTLENOSE DOLPHIN

The Bottlenose Dolphin varies greatly in size, shape, and color from one individual to another and according to the geographical region in which it lives; indeed, there may be several different species. However, there appear to be 2 main varieties: a smaller, inshore form, and a larger, more robust form that lives mainly offshore. Both have rather complex coloring although, under most light conditions at sea, they appear to be a uniform, quite featureless gray color. The main characteristics are its prominent, dark dorsal fin, combined with an inquisitive and active behavior. Confusion is possible with other gray dolphins, such as Tucuxis (p.172), Rough-toothed Dolphins (p.190), Risso's Dolphins (p.206), hump-backed dolphins (pp.174–177), and spotted dolphins (pp.184–189). It is fairly common and widespread, but population declines have been noted recently in parts of northern Europe, the Mediterranean, and the Black Sea.

• **OTHER NAMES** Gray Porpoise, Black Porpoise, Bottle-nosed Dolphin, Atlantic (or Pacific) Bottlenose Dolphin, Cowfish.

sharp crease between beak and forehead •

rounded forehead • *(variable)*

dark bluish gray or brownish gray cape (usually indistinct at a distance) •

center of dorsal fin may be paler than margins •

fairly short beak (length and thickness variable) •

dark stripe from eye to flipper •

TEETH $\frac{40-52}{36-48}$

• *robust body and head*

broad base •

• *moderately long, dark, slender flippers*

• *off-white, light gray, or pinkish underside (may be some spotting in older animals)*

• *pointed tips*

BEHAVIOR

Highly active at the surface: frequently lobtails, bow-rides, wake-rides, body-surfs, rides pressure waves of large whales, and breaches (sometimes leaping several meters high). May be found in association with a variety of other cetaceans, as well as sharks and sea turtles. Wild, lone individuals (usually males) sometimes seek out swimmers and small boats, remaining in the same area for years. Powerful swimmer. Dives rarely last longer than 3 to 4 minutes inshore, but sometimes longer offshore. Typically shows forehead when surfacing, but rarely the beak. In some areas, will chase fish out of the water, beaching itself before wriggling back. Groups may provide mutual assistance to one another; sometimes cooperate with local fishers.

COLOR VARIATIONS
Bottlenose Dolphins vary greatly in size, shape, and color: here are just a few of the possible variations.

| Group size 1–10 (inshore); 1–25 (offshore), up to 500 may occur offshore | Fin position Center |

Status Common	Population Unknown	Threats

ID CHECKLIST

- subdued gray coloring
- dark dorsal cape
- prominent, falcate fin
- robust head and body
- distinct beak with melon crease
- rounded forehead
- usually in small groups
- often bow-rides
- can be highly active

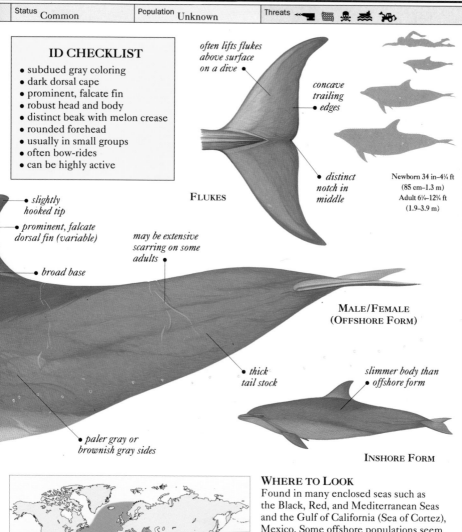

often lifts flukes above surface on a dive

concave trailing edges

FLUKES

distinct notch in middle

Newborn 34 in–4¼ ft
(85 cm–1.3 m)
Adult 6¼–12¾ ft
(1.9–3.9 m)

slightly hooked tip

prominent, falcate dorsal fin (variable)

may be extensive scarring on some adults

broad base

MALE/FEMALE (OFFSHORE FORM)

thick tail stock

slimmer body than offshore form

paler gray or brownish gray sides

INSHORE FORM

WIDELY DISTRIBUTED IN COLD TEMPERATE TO TROPICAL SEAS WORLDWIDE

WHERE TO LOOK

Found in many enclosed seas such as the Black, Red, and Mediterranean Seas and the Gulf of California (Sea of Cortez), Mexico. Some offshore populations seem to undertake seasonal migrations; many inshore populations are resident year-round. Outside tropical waters, seen mainly inshore in a wide range of coastal habitats from open coasts with strong surf to lagoons, large estuaries, and even the lower reaches of rivers and harbors. Offshore form common around oceanic islands, but can be seen in the open sea in the eastern tropical Pacific and elsewhere. In the northern North Atlantic, rare further north than the United Kingdom.

Birth wt 35–65 lb (15–30 kg)	Adult wt 330–1,435 lb (150–650 kg)	Diet

OCEANIC DOLPHINS
WITHOUT PROMINENT BEAKS

I N MANY PARTS of the world, oceanic dolphins are more likely to be encountered than any other cetaceans. Many of these dolphins are abundant, widespread, and highly visible: they are social animals (some species travel in schools of several thousand) and tend to be very active at the surface. Oceanic dolphins form a large family which in this book has been divided into 2 groups: those without prominent beaks (described here) and those with prominent beaks (pp.160–193); this classification is not generally recognized, but is simply intended to make identification easier.

smoothly sloping forehead (except Risso's and Irrawaddy Dolphins) •

• single blowhole

fairly prominent dorsal fin (variable) •

short, indistinct beak (longer in some Lagenorhynchus species) •

CHARACTERISTICS
There is great variation between the dolphins in this section. Besides the differences in color and pattern, the shapes of the bodies, beaks, flippers, and dorsal fins are also highly variable. Some species even change in appearance as they age. However, they do share a number of characteristics, including relatively short, indistinct beaks, fairly robust bodies, smoothly sloping foreheads (except Irrawaddy and Risso's Dolphins), and a notch in the middle of their flukes. Most have prominent dorsal fins.

• robust body

SKULL
A dead or stranded oceanic dolphin can be distinguished from a porpoise by its conical teeth (porpoises have spade-shaped teeth).

FRASER'S DOLPHIN

re-enters water head first

entire body comes out of water

begins to leave water at shallow angle

dolphin rises to surface at high speed

continues swimming at high speed

DIVE SEQUENCE (FAST-SWIMMING)
When swimming fast, many dolphins leap completely out of the water to breathe. This makes hydrodynamic sense, because it helps reduce turbulence and drag at the water surface, enabling them to maintain speed while using the minimum amount of energy.

LEAPING "DUSKY"
Most members of the genus Lagenorhynchus *(frequently referred to as "lags") are acrobatic; the Dusky Dolphin, in particular, is well known for its extraordinary high leaps and somersaults.*

tail stock variable
• from slim to stocky

• notch in
middle of flukes

COMMERSON'S DOLPHIN
Commerson's Dolphin has many features common to most members of the group, but it is unusual in having a predominantly light upper side: the other species have mainly dark upper sides and light undersides, camouflaging them from above (against the dark ocean depths) and from below (against the bright surface waters).

DORSAL FINS
All dolphins without prominent beaks have a dorsal fin. This normally has a concave trailing edge and is close to the center of the body but, nevertheless, there are many significant differences between species and individuals.

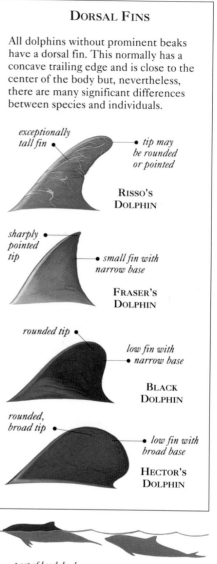

exceptionally
tall fin •

• tip may
be rounded
or pointed

**RISSO'S
DOLPHIN**

sharply
pointed
tip •

• small fin with
narrow base

**FRASER'S
DOLPHIN**

rounded tip •

low fin with
• narrow base

**BLACK
DOLPHIN**

rounded,
broad tip •

• low fin with
broad base

**HECTOR'S
DOLPHIN**

FRASER'S DOLPHIN

head dips and back
rolls forward

part of head, back,
and dorsal fin appear
above surface

dolphin rises
slowly to surface

dolphin disappears;
flukes rarely visible
when diving

DIVE SEQUENCE (SLOW-SWIMMING)
When swimming slowly, smaller oceanic dolphins without prominent beaks may resemble porpoises. As they rise to breathe, they barely disturb the surface of the water and show relatively little of themselves before disappearing again.

SPECIES IDENTIFICATION

COMMERSON'S DOLPHIN (p.198) *Striking black and white body pattern; body shape is rather like that of a porpoise, but conspicuous behavior is unmistakably dolphinlike.*

HOURGLASS DOLPHIN (p.216) *An inhabitant of remote Antarctic seas, with distinctive black and white coloration and prominent dorsal fin.*

HECTOR'S DOLPHIN (p.204) *One of the smallest cetaceans, with a distinctive rounded dorsal fin and complex body pattern of gray, black, and white; one of the rarest marine dolphins.*

DUSKY DOLPHIN (p.220) *One of the most acrobatic of all cetaceans, with a striking but complex body coloration; highly gregarious.*

HEAVISIDE'S DOLPHIN (p.202) *Poorly known dolphin with a robust body, prominent, triangular dorsal fin, and striking black, white, and gray coloration.*

PEALE'S DOLPHIN (p.214) *The dark face and brilliant white "armpits" help identify this fairly common but poorly known dolphin.*

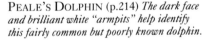

BLACK DOLPHIN (p.200) *Poorly known and inconspicuous dolphin, with a very limited distribution in southern Chile; large, rounded dorsal fin and stocky body.*

PACIFIC WHITE-SIDED DOLPHIN (p.218) *Highly active and demonstrative; very similar to the Dusky Dolphin, but ranges do not overlap.*

SPECIES IDENTIFICATION

ATLANTIC WHITE-SIDED DOLPHIN
(p.210) *A sociable dolphin with a distinctive yellowish streak on either side of its extremely thick tail stock; a fast swimmer and fairly acrobatic.*

FRASER'S DOLPHIN (p.208) *Not seen alive until the early 1970s, but there have been many sightings since then; can be identified by its stocky build, dark lateral stripe, and small dorsal fin.*

IRRAWADDY DOLPHIN (p.222)
Distinctive dolphin with a rounded head, small, stubby dorsal fin, and large, spatulate flippers; found in shallow coastal waters and rivers.

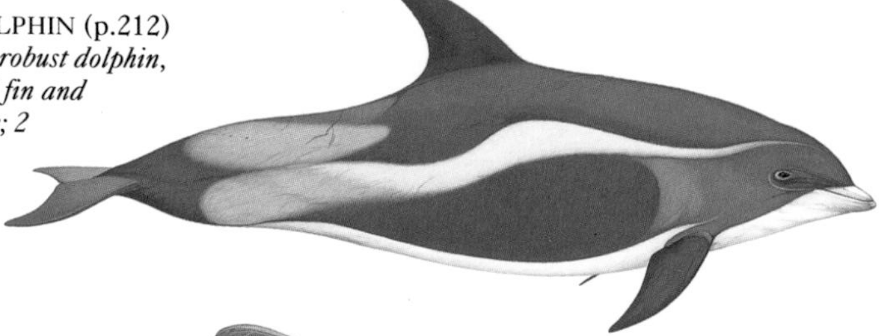

WHITE-BEAKED DOLPHIN (p.212)
Large and exceptionally robust dolphin, with a prominent dorsal fin and powerful swimming style; 2 white areas on flanks; may not necessarily have a white beak.

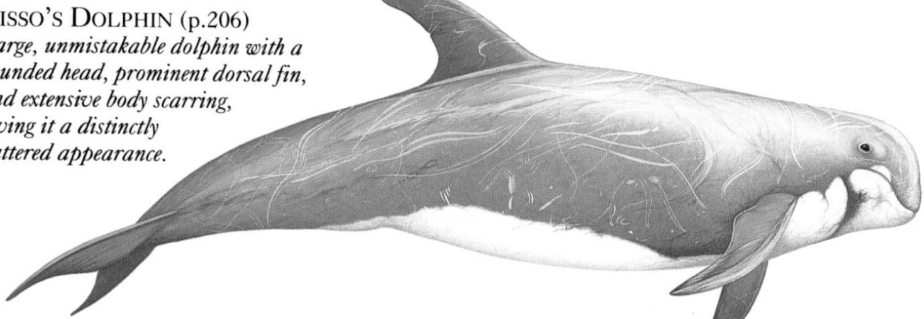

RISSO'S DOLPHIN (p.206)
Large, unmistakable dolphin with a rounded head, prominent dorsal fin, and extensive body scarring, giving it a distinctly battered appearance.

Family DELPHINIDAE	Species *Cephalorhynchus commersonii*	Habitat 〰〰

COMMERSON'S DOLPHIN

This is a striking animal and is relatively easy to identify at sea. Its small, stocky body is more similar to that of a porpoise than a dolphin, but its conspicuous behavior is unmistakably dolphinlike. At birth, it is gray, black, and brown; as it ages, it develops a subdued black and gray coloration, which turns starkly black and white when it is adult. There is considerable variation in appearance from one individual to another, particularly in the extent of black and white. It is possible to tell the sexes apart by the black patch on the underside, which is shaped like a raindrop in males and like a horseshoe in females. The population around Kerguelen Island, in

the Indian Ocean, is geographically isolated and may form a separate subspecies; most individuals are larger than the South American animals and are black, gray, and white. Hunting of Commerson's Dolphins in Chile and Argentina, mainly for use as bait in crab fisheries, may be a serious threat.

• **OTHER NAMES** Skunk Dolphin, Piebald Dolphin, Black-and-white Dolphin, Jacobite, Puffing Pig.

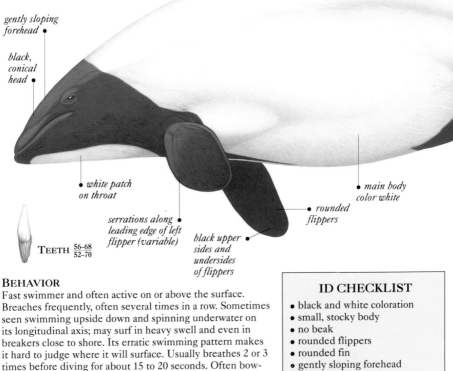

gently sloping forehead •

black, conical head •

• *white patch on throat*

serrations along • leading edge of left flipper (variable)

black upper • sides and undersides of flippers

• *main body color white*

• *rounded flippers*

TEETH $\frac{56-68}{52-70}$

BEHAVIOR
Fast swimmer and often active on or above the surface. Breaches frequently, often several times in a row. Sometimes seen swimming upside down and spinning underwater on its longitudinal axis; may surf in heavy swell and even in breakers close to shore. Its erratic swimming pattern makes it hard to judge where it will surface. Usually breathes 2 or 3 times before diving for about 15 to 20 seconds. Often bow-rides and will swim alongside or behind vessels. Sometimes found with Peale's and Black Dolphins and Burmeister's Porpoises. Some populations may keep to quite well-defined territories. Probably forages on or close to sea floor.

ID CHECKLIST
• black and white coloration
• small, stocky body
• no beak
• rounded flippers
• rounded fin
• gently sloping forehead
• black flippers, flukes, and fin
• usually in small groups
• likely to approach boats

Group size 1–3 (1–15), occasionally in groups of 100 or more	Fin position Slightly behind center

Status Locally common	Population Unknown	Threats 

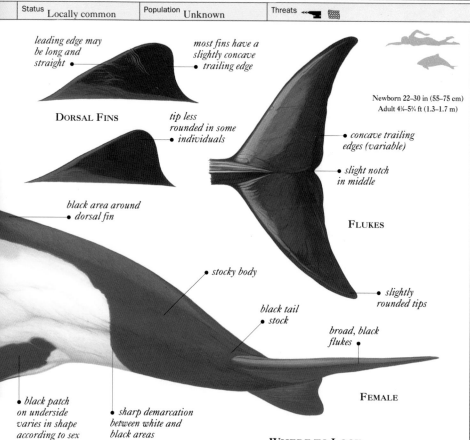

leading edge may be long and straight •

most fins have a slightly concave trailing edge •

Newborn 22–30 in (55–75 cm)
Adult 4¼–5¾ ft (1.3–1.7 m)

DORSAL FINS

tip less rounded in some individuals •

• concave trailing edges (variable)

• slight notch in middle

black area around dorsal fin •

FLUKES

• stocky body

black tail stock •

• slightly rounded tips

broad, black flukes •

FEMALE

• black patch on underside varies in shape according to sex

• sharp demarcation between white and black areas

SOUTHERN SOUTH AMERICA, INCLUDING THE FALKLAND ISLANDS, AND KERGUELEN ISLAND IN THE INDIAN OCEAN

WHERE TO LOOK

Distribution appears to be continuous along the South American coast from Peninsula Valdés, Argentina, to Tierra del Fuego. Also occurs in Chilean waters south of 51° S and around the Falkland Islands and Kerguelen Island, and there are scattered records from waters south of Tierra del Fuego. An earlier record from South Georgia is unreliable. Appears to be most common in southern Tierra del Fuego, around the Falklands (especially near harbors and natural protected areas), and in the Straits of Magellan. Most sightings are close to shore, in water less than 325 ft (100 m) deep. Found along open coasts and in fjords, bays, and river mouths; known to enter rivers. Seems to prefer areas with a large tidal range. Often near kelp beds.

Birth wt c.13 lb (6 kg)	Adult wt 75–130 lb (35–60 kg)	Diet

Family DELPHINIDAE	Species *Cephalorhynchus eutropia*	Habitat 〰〰

BLACK DOLPHIN

The Black Dolphin is one of the smallest of all cetaceans. It is very poorly known, from a mixed collection of skeletons, a handful of strandings, and a limited number of sightings. Confusion is most likely to occur with the Spectacled Porpoise (p.240) in the south of its range and Burmeister's Porpoise (p.246) in the north; the dorsal fin shape is a good distinguishing feature of all three species. There is also some overlap with the black

and white Commerson's Dolphin (p.198) in the extreme south. The Black Dolphin is hunted illegally to provide bait for use in traps in the king crab fishery in Chile – a major cause for concern, since the dolphin's population could be very low. The body color darkens very quickly after death, possibly accounting for inaccurate descriptions in early reports.
• **OTHER NAMES** White-bellied Dolphin, Chilean Dolphin, Chilean Black Dolphin.

pale gray "cap" on forehead

conical head with gently sloping forehead

indistinct beak

fairly large, rounded dorsal fin

white "lips"

white throat

small, rounded flippers

small white patch behind each flipper

white belly

BEHAVIOR

Little information on behavior, but it is generally thought to be unobtrusive. Rarely breaches. Reported to have a slightly undulating movement in the water, rather like a swimming sea lion. Often seen among breakers and swells very close to shore. Animals in southern part of range tend to be more wary of boats and difficult to approach; in the north, they have been known to swim over to boats and may bow-ride. Groups tend to be far larger along the open coast in the north, and as many as 4,000 animals have been seen traveling together. Often found in association with flocks of feeding seabirds.

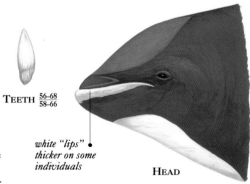

TEETH $\frac{56-68}{58-66}$

white "lips" thicker on some individuals

HEAD

Group size 2–3 (2–10), larger temporary gatherings	Fin position Center

Status Rare	Population Unknown	Threats 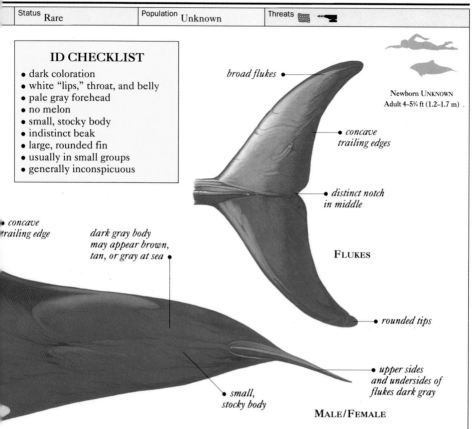

ID CHECKLIST
- dark coloration
- white "lips," throat, and belly
- pale gray forehead
- no melon
- small, stocky body
- indistinct beak
- large, rounded fin
- usually in small groups
- generally inconspicuous

broad flukes •

Newborn UNKNOWN
Adult 4–5¾ ft (1.2–1.7 m)

• *concave
trailing edges*

• *distinct notch
in middle*

FLUKES

• *concave
trailing edge*

*dark gray body
may appear brown,
tan, or gray at sea* •

• *rounded tips*

• *upper sides
and undersides of
flukes dark gray*

• *small,
stocky body*

MALE/FEMALE

WHERE TO LOOK
Restricted to cold, shallow, coastal waters of Chile. Range
stretches from Valparaíso in the north to Navarino Island,
near Cape Horn, in the extreme south. Also occurs in the
Straits of Magellan and the channels of Tierra del Fuego.
Distribution seems to be continuous, though there seem
to be areas of local abundance, such as off Playa Frailes,
Valdivia, Golfo de Arauco, and near Isla de Chiloé.
Known to enter Rio Valdivia and other rivers. May also
sometimes occur at the extreme southern tip of Argentina.
No seasonal movements have been recognized. Seems
to prefer areas with significant tidal range. Frequently
encountered at entrances to fjords and in bays and
river mouths, but also occurs along fairly open coast.
There is no information on the distance ranged offshore.

COASTAL WATERS OF CHILE

Birth wt Unknown	Adult wt c.65–145 lb (30–65 kg)	Diet

| Family DELPHINIDAE | Species *Cephalorhynchus heavisidii* | Habitat |

HEAVISIDE'S DOLPHIN

Heaviside's Dolphin is a poorly known species, rarely seen in the wild. Information on live animals has recently been published for the first time, and previous illustrations (from dead animals) are now known to be incorrect. It does not resemble any other species off the coast of southwestern Africa, so should be relatively easy to identify. It is a small, compact dolphin, with a robust shape typical of the genus *Cephalorhynchus* and striking coloration. Some Heaviside's become entangled in a variety of inshore fishing nets off South Africa and Namibia each year, and small numbers may be taken with hand-held harpoons or rifles for human consumption.

• **OTHER NAMES** South African Dolphin, Benguela Dolphin.

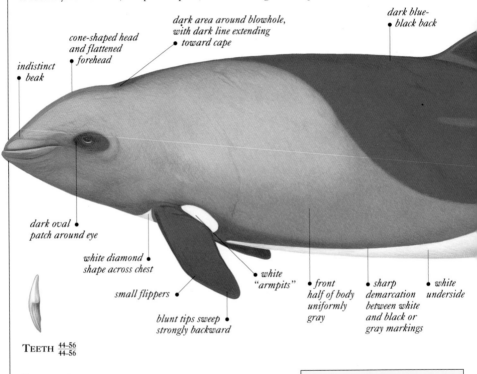

indistinct beak

cone-shaped head and flattened forehead

dark area around blowhole, with dark line extending toward cape

dark blue-black back

dark oval patch around eye

white diamond shape across chest

small flippers

blunt tips sweep strongly backward

white "armpits"

front half of body uniformly gray

sharp demarcation between white and black or gray markings

white underside

TEETH $\frac{44-56}{44-56}$

BEHAVIOR
Little is known about this species' behavior. It is generally undemonstrative and appears to be shy. Breaching is rare, but it is known to leap more than 6½ ft (2 m) high. Has been observed doing rapid forward somersaults, which end in a tail-slap on the surface. May porpoise when swimming at high speed. Reaction to vessels varies, but is known to approach a range of boats and to bow-ride and wake-ride; some animals have been seen "escorting" small vessels for several hours at a time. Limited observations suggest that at least some groups have restricted home ranges and probably do not stray far from these areas.

ID CHECKLIST
• prominent triangular fin
• small, stocky body
• gray anterior, dark posterior
• white belly with "lobes"
• gray, cone-shaped head
• indistinct beak
• dark flippers, fin, and flukes
• generally in small groups
• usually undemonstrative

| Group size 2–3 (1–10), temporary gatherings of up to 30 | Fin position Slightly behind center |

Status Rare	Population Unknown	Threats 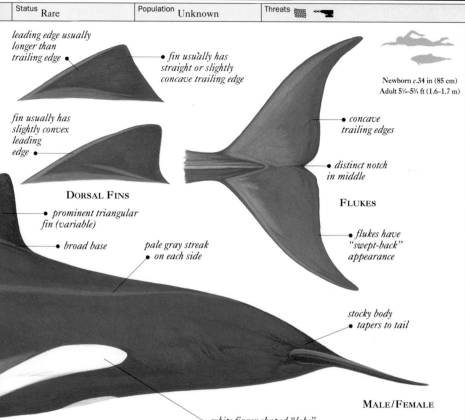

leading edge usually
longer than
trailing edge •

• fin usually has
straight or slightly
concave trailing edge

Newborn *c.*34 in (85 cm)
Adult 5¼–5¾ ft (1.6–1.7 m)

*fin usually has
slightly convex
leading
edge •*

• *concave
trailing edges*

DORSAL FINS

• *distinct notch
in middle*

FLUKES

• *prominent triangular
fin (variable)*

• *broad base*   *pale gray streak
• on each side*

• *flukes have
"swept-back"
appearance*

*stocky body
• tapers to tail*

MALE/FEMALE

• *white finger-shaped "lobe"
points toward tail*

WHERE TO LOOK

Range is restricted and fairly sparsely populated
throughout. Occurs only off western coasts of South Africa
and Namibia, along approximately 1,000 miles (1,600 km)
of shoreline. Distribution known from Cape of Good Hope,
South Africa, north to at least Cape Cross, Namibia, but
may extend as far north as southern Angola. Mostly seen in
coastal waters, within 5–6 miles (8–10 km) of shore and in
water less than 330 ft (100 m) deep. Surveys within 5 miles
(8 km) of the coast have shown low population densities of
around 5 sightings per 100 miles (160 km); sightings
dropped dramatically farther offshore, and no animals were
seen in water deeper than 655 ft (200 m). Seems to be
associated with the cold, northward-flowing Benguela
Current. Some populations may be resident year-round.

COASTAL WATERS FROM SOUTHERN SOUTH AFRICA
NORTHWARD TO CENTRAL NAMIBIA

Birth wt Unknown	Adult wt 90–165 lb (40–75 kg)	Diet

| Family DELPHINIDAE | Species *Cephalorhynchus hectori* | Habitat 〰️ |

HECTOR'S DOLPHIN

Hector's Dolphin is one of the smallest of all cetaceans: most individuals are less than 4¾ ft (1.4 m) long. It can appear dark from a distance, but at close range has a striking, complex color pattern of gray, black, and white. The rounded dorsal fin, with its convex trailing edge, makes identification at sea relatively easy. Males are slightly smaller than females and have a large dark gray patch around the genital slit. Hector's Dolphin is a familiar sight in certain locations but is actually one of the rarest marine dolphins in the world. It does not appear to be in immediate danger, but incidental catches in coastal gill nets give cause for concern. During a study around Banks Peninsula, on South Island, New Zealand, nearly a third of an estimated 760 dolphins drowned in nets during the period 1984 to 1988; fortunately, the area has since been declared a marine sanctuary.

• OTHER NAMES Little Pied Dolphin, New Zealand Dolphin, New Zealand White-front Dolphin.

gray forehead
thinly streaked
• with black

• tip of
indistinct
beak black

• white throat
and chest

dark patch •
from eye area
to flipper

TEETH $\frac{52-64}{52-64}$

• small white patch
behind each flipper

• large, dark,
rounded flippers

• white belly
with dark
border

BEHAVIOR
Rarely bow-rides, but frequently swims in the wake of passing boats; may also swim alongside boats for short distances. Unlike many dolphins, it prefers stationary or slow vessels (less than 10 knots) and will dive to avoid faster ones. Inquisitive. Sometimes breaches (usually re-entering the water without out a splash) and may lobtail, spyhop, and surf. Surfaces frequently to breathe, showing little of itself and, especially on calm days, barely making a splash. May rest motionless at the surface. Groups rarely stay in tight formation, though several individuals may swim and surface together in a row. Most active when small groups join together.

ID CHECKLIST
• rounded fin
• no discernible beak
• pale but complex color pattern
• pale gray forehead
• white, finger-shaped lobe
• small size
• dark flippers, fin, and flukes
• barely disturbs water
• usually in small groups

| Group size 2–8 (2–30), loose aggregations of 100 or more in some areas | Fin position Slightly behind center |

Status Endangered	Population 3,000–4,000	Threats

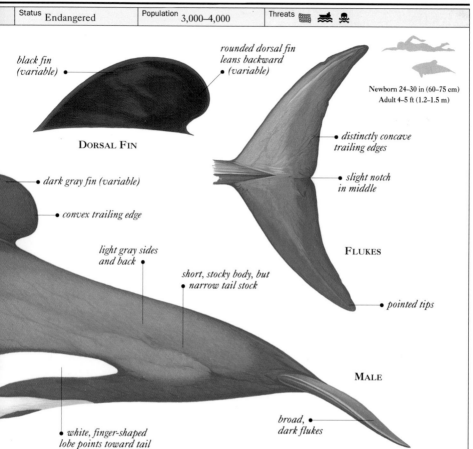

black fin (variable)

rounded dorsal fin leans backward (variable)

Newborn 24–30 in (60–75 cm)
Adult 4–5 ft (1.2–1.5 m)

DORSAL FIN

• distinctly concave trailing edges

• dark gray fin (variable)

• slight notch in middle

• convex trailing edge

light gray sides and back •

FLUKES

short, stocky body, but narrow tail stock •

• pointed tips

MALE

broad, • dark flukes

• white, finger-shaped lobe points toward tail

WHERE TO LOOK

Lives exclusively around New Zealand. Most common around South Island, particularly Banks Peninsula and Cloudy Bay, and along the western coast of North Island, mainly between Kawhia and Manukau Harbor. Tends to be abundant in patches and is absent from some areas within the range. May make small movements inshore in summer and offshore in winter. Best place to look is in shallow water, along rocks close to the shore. May enter estuaries, and known to swim short distances up river. Usually within ⅝ mile (1 km) of the shoreline and rarely farther than 5 miles (8 km). Earlier reports from Australia and Sarawak, Malaysia, were incorrectly identified.

COASTAL WATERS OF NEW ZEALAND, ESPECIALLY SOUTH ISLAND AND THE WESTERN COAST OF NORTH ISLAND

Birth wt c.20 lb (9 kg)	Adult wt 75–130 lb (35–60 kg)	Diet

Family DELPHINIDAE	Species *Grampus griseus*	Habitat 〰️ 🐋

RISSO'S DOLPHIN

Risso's Dolphins are relatively easy to identify at sea, particularly when they are older. They develop a distinctly battered appearance, with extensive body scarring caused by the teeth of other Risso's Dolphins and, to a lesser extent, by confrontations with squid. Their body color also tends to lighten with age, although there is a great deal of variation between individuals: adults may be almost as white as Belugas (p.92) or as dark as pilot whales (pp.148–151). From a distance, the tall dorsal fin may cause momentary confusion with female or juvenile Killer Whales (p.152) or Bottlenose Dolphins (p.192). Risso's Dolphin has a crease down the center of the forehead, from the blowhole to the upper "lip"; this is visible at close range and is unique to this species. Risso's Dolphin is sometimes seen in mixed schools with several species of dolphin and with pilot whales.

• OTHER NAMES Gray Dolphin, White-head Grampus, Gray Grampus, Grampus.

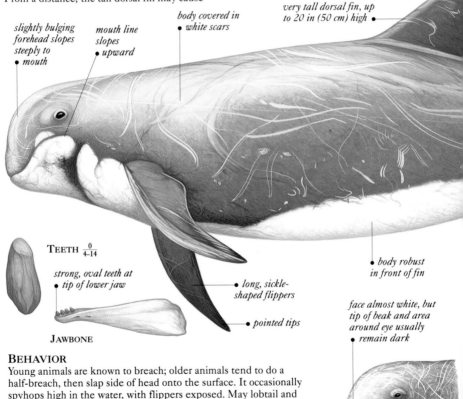

very tall dorsal fin, up to 20 in (50 cm) high •

slightly bulging forehead slopes steeply to • *mouth*

mouth line slopes • *upward*

body covered in • *white scars*

TEETH $\frac{0}{4-14}$

strong, oval teeth at • *tip of lower jaw*

JAWBONE

long, sickle-shaped flippers •

pointed tips •

• *body robust in front of fin*

face almost white, but tip of beak and area around eye usually • *remain dark*

OLD ANIMAL

BEHAVIOR

Young animals are known to breach; older animals tend to do a half-breach, then slap side of head onto the surface. It occasionally spyhops high in the water, with flippers exposed. May lobtail and flipper-slap, and will surf in waves. Seldom bow-rides but may swim alongside a vessel or in its wake. Typically dives for 1 to 2 minutes, then takes up to a dozen breaths at 15- to 20-second intervals; can stay underwater for up to 30 minutes. Flukes may appear above water when diving. Sometimes swims by porpoising. May surface at a 45° angle to breathe. Groups sometimes spread out in a long line when hunting. Some groups very shy, but others allow a close approach.

Group size 3–50 (1–150), temporary gatherings of several hundred	Fin position Center

Status Common	Population Unknown	Threats

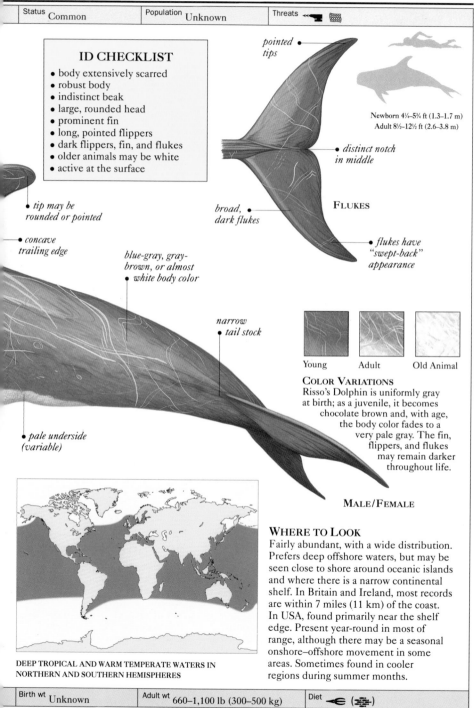

ID CHECKLIST

- body extensively scarred
- robust body
- indistinct beak
- large, rounded head
- prominent fin
- long, pointed flippers
- dark flippers, fin, and flukes
- older animals may be white
- active at the surface

• pointed tips

Newborn 4¼–5¾ ft (1.3–1.7 m)
Adult 8½–12½ ft (2.6–3.8 m)

• distinct notch in middle

• tip may be rounded or pointed

broad, • dark flukes

FLUKES

• concave trailing edge

blue-gray, gray-brown, or almost • white body color

• flukes have "swept-back" appearance

narrow • tail stock

Young   Adult   Old Animal

COLOR VARIATIONS
Risso's Dolphin is uniformly gray
at birth; as a juvenile, it becomes
chocolate brown and, with age,
the body color fades to a
very pale gray. The fin,
flippers, and flukes
may remain darker
throughout life.

• pale underside
(variable)

MALE/FEMALE

WHERE TO LOOK
Fairly abundant, with a wide distribution.
Prefers deep offshore waters, but may be
seen close to shore around oceanic islands
and where there is a narrow continental
shelf. In Britain and Ireland, most records
are within 7 miles (11 km) of the coast.
In USA, found primarily near the shelf
edge. Present year-round in most of
range, although there may be a seasonal
onshore–offshore movement in some
areas. Sometimes found in cooler
regions during summer months.

DEEP TROPICAL AND WARM TEMPERATE WATERS IN
NORTHERN AND SOUTHERN HEMISPHERES

Birth wt Unknown	Adult wt 660–1,100 lb (300–500 kg)	Diet

Family DELPHINIDAE	Species *Lagenodelphis hosei*	Habitat 〰〰

FRASER'S DOLPHIN

Although a carcass of Fraser's Dolphin was found on a beach in Sarawak, Malaysia, in 1895, the species was not scientifically described until 1956 and was not seen alive until the early 1970s. There have been many observations at sea since then, and it does not appear to be as rare as was once thought. However, it is still relatively poorly known. It is intermediate in appearance between the genera *Lagenorhynchus* and *Delphinus*, thus the composite generic name *Lagenodelphis*. Some individuals, especially males, have a very striking, black lateral body stripe; the width and intensity of the stripe are believed to increase with age. There may be some confusion with the Striped Dolphin (p.178), though Fraser's Dolphin has a shorter beak, a smaller dorsal fin, and tiny flippers, as well as different body striping. Unknown numbers are drowned in pelagic drift-net fisheries and other fishing operations, and there is some direct hunting within the range.

• **OTHER NAMES** Sarawak Dolphin, Shortsnout Dolphin, Bornean Dolphin, White-bellied Dolphin, Fraser's Porpoise.

blue-gray or gray-
• brown upper side

dorsal fin small •
in relation to
body size

complex face
• markings

short but well-
• defined beak

• upper jaw
and tip of
lower jaw
dark

• dark line
(or lines)
from beak
to flippers

TEETH $\frac{72-88}{68-88}$

• flippers
dark on
both sides

• very small,
pointed flippers

• creamy white or
pinkish white belly
and throat

• gray or creamy line
bordering upper side
of dark stripe

BEHAVIOR

Analysis of prey suggests Fraser's Dolphin is a deep diver, hunting at depths of at least 820–1,640 ft (250–500 m). Often seen in mixed schools with other pelagic cetaceans, especially Melon-headed, False Killer, and Sperm Whales, and Pantropical Spotted and Striped Dolphins. Has an aggressive swimming style: when rising to breathe, often leaves water in a burst of spray. Known to breach, though not usually demonstrative or playful. In most parts of range is shy of boats and will normally swim away rapidly, forming a tight group with others, with lots of surface splashing. More inclined to bow-ride and swim alongside vessels in the Philippines and off Natal coast, South Africa.

ID CHECKLIST

- stocky build
- dark lateral stripe
- small fin
- short beak
- tiny flippers
- aggressive swimming style
- frequently in large groups
- often in mixed schools
- shy of boats in most areas

Group size 100–500 (4–1,000), often in the company of other species	Fin position Center

Status Locally common	Population Unknown	Threats

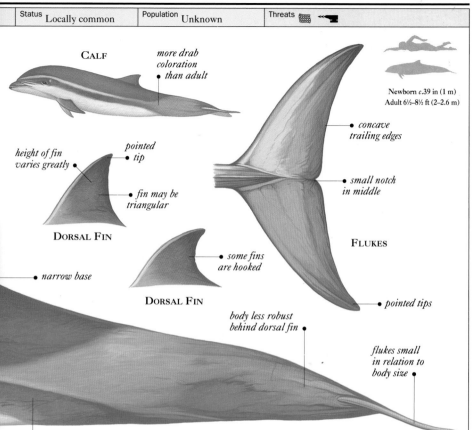

CALF

more drab
coloration
• than adult

Newborn c.39 in (1 m)
Adult 6½–8½ ft (2–2.6 m)

• concave
trailing edges

height of fin
varies greatly •

pointed
• tip

• small notch
in middle

• fin may be
triangular

DORSAL FIN

FLUKES

• some fins
are hooked

• narrow base

DORSAL FIN

• pointed tips

body less robust
behind dorsal fin •

flukes small
in relation to
body size •

• dark gray to black lateral stripe
(variable width and intensity)

MALE/FEMALE

WHERE TO LOOK

Distribution poorly known. Appears to be most common near equator in the eastern tropical Pacific and at southern end of Bohol Strait, the Philippines. Seems to be relatively scarce in the Atlantic Ocean (known only from the Lesser Antilles and the Gulf of Mexico). May range across the Indian Ocean, though confirmed sightings only from the east coast of South Africa, Madagascar, Sri Lanka, and Indonesia. Also occurs away from equator as far north as Taiwan and Japan and, in small numbers, off Australia. A stranding in France was probably a vagrant school. May be more common than relative lack of records implies. Rarely seen in inshore waters, except around oceanic islands and in areas with a narrow continental shelf.

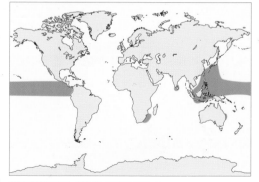

DEEP TROPICAL AND WARM TEMPERATE WATERS OF THE
PACIFIC, ATLANTIC, AND INDIAN OCEANS

Birth wt c.40 lb (19 kg)	Adult wt c.350–460 lb (160–210 kg)	Diet

Family DELPHINIDAE	Species *Lagenorhynchus acutus*	Habitat 〰〰 ▨▨

ATLANTIC WHITE-SIDED DOLPHIN

The Atlantic White-sided Dolphin is fairly large and robust and is very conspicuous at sea. A sociable animal, it is often in the company of White-beaked Dolphins, Humpback Whales, Fin Whales, and Long-finned Pilot Whales. Confusion is most likely with the White-beaked Dolphin (p.212); however, the Atlantic White-Sided Dolphin is slightly smaller and slimmer; it also has a white patch on both sides, below the dorsal fin, running into a yellowish streak on either side of the tail stock. There may be confusion with the Common Dolphin (p.164)

as well, because it has a superficially similar pattern of gray, white, black, and yellow; however, the Atlantic White-sided Dolphin has a stockier body, a shorter beak, and no distinctive hourglass pattern on its sides.

• **OTHER NAMES** Jumper, Springer, Lag, Atlantic White-sided Porpoise.

tall, falcate dorsal fin •
(more erect in adult males)

uniformly black •
or dark gray
dorsal fin

dark ring •
around eye

gently sloping
forehead •

• beak black or dark
gray above, white or
pale gray below

TEETH $\frac{58-80}{58-80}$

• dark
stripe between
corner of mouth
and flipper
(variable)

• black or dark
gray sickle-
shaped flippers

• pointed tips

• white band
below dorsal fin

yellow and white
patches usually visible
simultaneously when it
surfaces to breathe
• surfaces to breathe

ATLANTIC WHITE-SIDED DOLPHIN

grayish patch visible
when it surfaces to
• breathe

BEHAVIOR
Acrobatic and a fast swimmer. Frequently breaches (though not as often as White-beaked or Common Dolphins) and lobtails. Surfaces to breathe every 10 to 15 seconds, either leaping clear of the water or barely breaking the surface and creating a wave over its head. Wary of ships in some areas, but will swim alongside slower vessels and may bow-ride in front of faster ones; sometimes rides the bow waves of large whales. Generally found in larger schools offshore and smaller ones inshore. Individual and mass strandings are relatively common.

WHITE-BEAKED DOLPHIN

Group size 5–50 (1–100), schools of up to 1,000 recorded offshore	Fin position Slightly forward of center

Status Locally common	Population Unknown	Threats 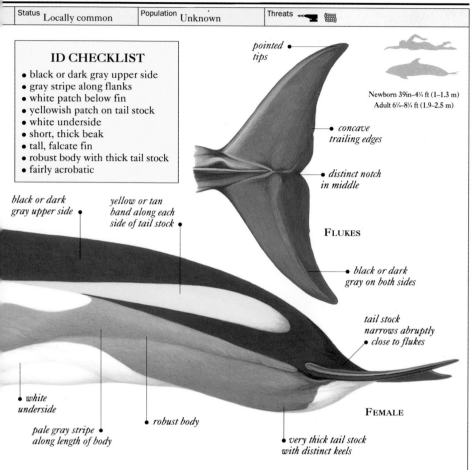

ID CHECKLIST

- black or dark gray upper side
- gray stripe along flanks
- white patch below fin
- yellowish patch on tail stock
- white underside
- short, thick beak
- tall, falcate fin
- robust body with thick tail stock
- fairly acrobatic

pointed •
tips

Newborn 39in–4¼ ft (1–1.3 m)
Adult 6¼–8¼ ft (1.9–2.5 m)

• concave
trailing edges

• distinct notch
in middle

FLUKES

black or dark
gray upper side •

yellow or tan
band along each
side of tail stock •

• black or dark
gray on both sides

tail stock
narrows abruptly
• close to flukes

• white
underside

pale gray stripe •
along length of body

• robust body

FEMALE

• very thick tail stock
with distinct keels

COOL TEMPERATE AND SUBARCTIC WATERS OF THE
NORTHERN NORTH ATLANTIC

WHERE TO LOOK

Range very similar to that of the White-beaked Dolphin (p.212). Toward the east of the range, may occasionally be found as far north as the southern Barents Sea and rarely seen farther south than the English Channel. In the west, has been reported from west Greenland to Chesapeake Bay (though usually from Cape Cod, northward); appears to be especially abundant in the Gulf of Maine, and large schools penetrate far up the St. Lawrence estuary, Canada. May be an inshore–offshore movement with the seasons in some areas. Seems to prefer areas with high sea floor relief and along the edge of the continental shelf.

Birth wt 65–75 lb (30–35 kg)	Adult wt 365–440 lb (165–200 kg)	Diet

Family DELPHINIDAE	Species *Lagenorhynchus albirostris*	Habitat

WHITE-BEAKED DOLPHIN

The White-beaked Dolphin is a large and very robust dolphin. Its common name is a little misleading, as the beak is not always white, but genuinely white-beaked individuals are distinctive at close range. Toward the east of the range, animals tend to have white beaks and live in smaller schools, while western animals generally have darker beaks and live in larger schools (though there are exceptions). The body pattern of white, gray, and black is also highly variable between individuals. Confusion is most likely with the Atlantic White-sided Dolphin (p.210); however, the White-beaked is slightly larger and more robust, and does not have the yellow-streaked sides characteristic of the Atlantic White-sided Dolphin.

• **OTHER NAMES** White-nosed Dolphin, Squidhound, White-beaked Porpoise.

tall, falcate dorsal fin, especially in adult males •

black dorsal fin with broad base •

mainly dark upper side •

white beak • (variable)

short, thick beak •

solid line from flippers to corner of mouth •

broad base •

medium-sized black flippers •

robust body •

pointed tips •

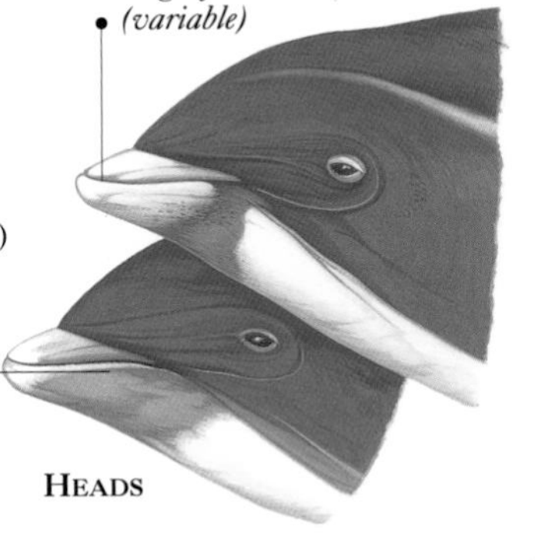

dark gray beak • (variable)

TEETH $\frac{44-56}{44-56}$

BEHAVIOR

May bow-ride, especially in front of large, fast-moving vessels, but usually loses interest quickly. However, some populations are very elusive. Sometimes acrobatic (especially when feeding) and will breach, normally falling onto its side or back. Typically a fast, powerful swimmer and in some parts of range may create a "rooster tail" reminiscent of Dall's Porpoise (p.248). When swimming at speed, may lift its whole body out of the water briefly as it rises to breathe. Has been seen with Fin Whales and Killer Whales, and may mix with other species.

mottled • brown beak (variable)

HEADS

Group size 2–30 (1–50), aggregations of 1,500 have been reported	Fin position Center

Status Common	Population Unknown	Threats 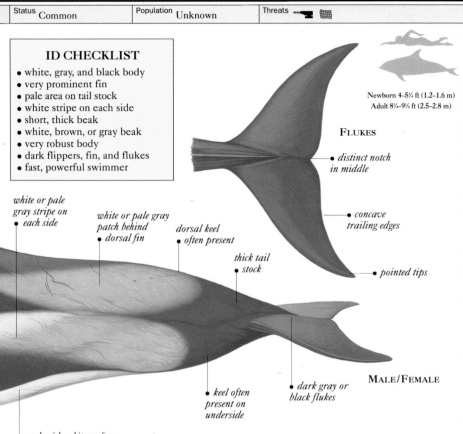

ID CHECKLIST

- white, gray, and black body
- very prominent fin
- pale area on tail stock
- white stripe on each side
- short, thick beak
- white, brown, or gray beak
- very robust body
- dark flippers, fin, and flukes
- fast, powerful swimmer

Newborn 4–5¼ ft (1.2–1.6 m)
Adult 8¼–9¼ ft (2.5–2.8 m)

FLUKES

- *distinct notch in middle*

white or pale gray stripe on each side

white or pale gray patch behind dorsal fin

dorsal keel often present

thick tail stock

- *concave trailing edges*

- *pointed tips*

MALE/FEMALE

- *keel often present on underside*

- *dark gray or black flukes*

- *underside white as far as middle of tail stock*

WHERE TO LOOK

This is the most northerly member of the genus *Lagenorhynchus*, and has a wide distribution. Animals in northernmost part of the range occur right up to the edge of the pack ice. Southern limit in west of the range is around Cape Cod; east of the range, animals occur as far south as Portugal but are rarely seen south of Britain. In some areas, there may be either an inshore–offshore movement or a north–south movement with the seasons (wintering in the south or offshore); in other areas, such as Britain, appears to be present all year round (but with seasonal peaks of abundance in coastal waters). Found widely over the continental shelf, but especially along the shelf edge.

COOL TEMPERATE AND SUBARCTIC WATERS OF THE NORTH ATLANTIC

Birth wt 90 lb (40 kg)	Adult wt 395–605 lb (180–275 kg)	Diet

| Family DELPHINIDAE | Species *Lagenorhynchus australis* | Habitat 〰 |

PEALE'S DOLPHIN

Peale's Dolphin is a fairly common animal around the southern tip of South America, although its isolated range means that it is rarely seen and quite poorly known. It is relatively easy to identify at sea, but may be confused with the Dusky Dolphin (p.220); however, unlike the Dusky, Peale's has a dark face and chin, a predominantly dark dorsal fin, white "armpits," and only a single grayish white body stripe on each side. Confusion may also occur with the Hourglass Dolphin (p.216). There is considerable concern about unknown numbers of Peale's Dolphins that become accidentally entangled in fishing nets and are hunted with harpoons; the meat is used as bait in crab traps.
• **OTHER NAMES** Blackchin Dolphin, Peale's Black-chinned Dolphin, Southern Dolphin, Peale's Porpoise.

large, predominantly grayish black dorsal fin •

broad base •

dark patch around each eye •

grayish black face and chin •

gently sloping forehead •

short, indistinct beak •

dark line separates white chest and belly from grayish white sides •

brilliant white patch at "armpits" •

grayish white sides •

some individuals have paler patch around each eye •

small, pointed flippers with convex leading edges •

TEETH $\frac{54-66}{54-66}$

BEHAVIOR
Known to ride bow waves of large vessels and may swim alongside smaller ones. Sometimes swims slowly but can be energetic and acrobatic, frequently leaping high into the air and falling back into the water, on its side, with a splash. May travel with long, low leaps. Limited evidence suggests it keeps to a specific and rather small home range. Has been observed playing in surf in the company of Risso's Dolphins. Food and feeding habits unknown, although an individual collected in the Falkland Islands had the remains of an octopus in its stomach; may also eat fish and squid.

HEAD

extent of darkness on chin variable •

| Group size 3–8 (1–30), many groups observed in temporary aggregations | Fin position Center |

Status Locally common	Population Unknown	Threats 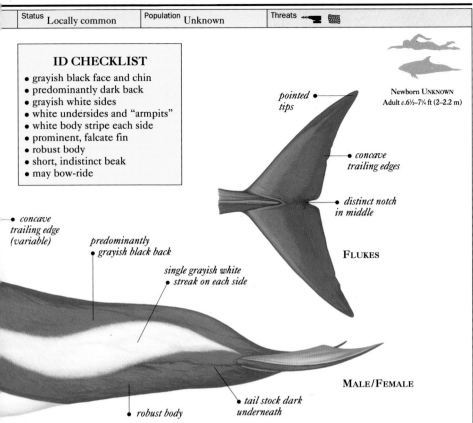

ID CHECKLIST

- grayish black face and chin
- predominantly dark back
- grayish white sides
- white undersides and "armpits"
- white body stripe each side
- prominent, falcate fin
- robust body
- short, indistinct beak
- may bow-ride

pointed • *tips*

Newborn UNKNOWN
Adult *c*.6½–7¼ ft (2–2.2 m)

• *concave trailing edges*

• *distinct notch in middle*

FLUKES

• *concave trailing edge (variable)*

predominantly • *grayish black back*

single grayish white • *streak on each side*

MALE/FEMALE

• *tail stock dark underneath*

• *robust body*

WHERE TO LOOK

Range known to extend from Golfo San Matías, Argentina, around tip of South America to Valparaíso, Chile (though most common south of Puerto Montt, Chile); may occur farther north in both countries. Recorded as far south as 57° S. Particularly common around the Falkland Islands and Tierra del Fuego (especially the Straits of Magellan and Beagle Channel); one of the most frequently sighted cetacean species in the Straits of Magellan. Distribution may be continuous between Argentina and the Falklands. Possible sighting at Palmerston Atoll, in the South Pacific, has not been confirmed. Frequently seen close to shore, in fjords, bays, and inlets (especially near kelp beds), but also over the continental shelf. Appears to have been marked decrease in number of sightings in areas of extreme south, where crab fishing takes place.

COOL, COASTAL WATERS OF SOUTHERN SOUTH AMERICA, INCLUDING THE FALKLAND ISLANDS

Birth wt Unknown	Adult wt *c*.255 lb (115 kg)	Diet Unknown

Family DELPHINIDAE	Species *Lagenorhynchus cruciger*	Habitat 〰〰

HOURGLASS DOLPHIN

An inhabitant of remote Antarctic and sub-antarctic seas, the Hourglass Dolphin was first described as long ago as 1824, but it is rarely seen and is still poorly known. With a concerted research effort, there have been more sightings in recent years, and it is commonly encountered during survey work in out-of-the-way waters rarely frequented by other vessels. It is a relatively easy animal to identify, as it has striking black and white body markings – it was named for the crude white hourglass pattern on its flanks – and because it is the only dolphin with a dorsal fin that is consistently found in southern polar waters. In the north of its range, the main sources of confusion are likely to be Peale's Dolphin (p.214) and Dusky Dolphin (p.220). The body color darkens dramatically soon after death.
• **OTHER NAMES** Wilson's Dolphin, Southern White-sided Dolphin.

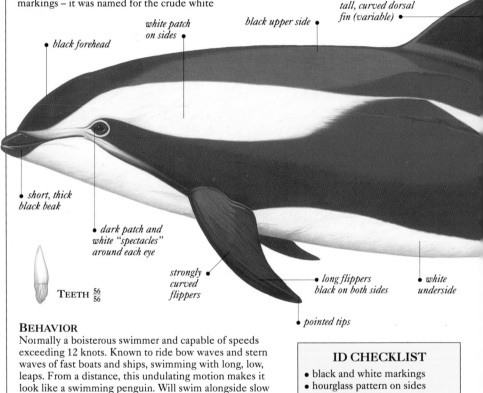

white patch on sides •

black forehead •

black upper side •

tall, curved dorsal fin (variable) •

short, thick black beak •

dark patch and white "spectacles" around each eye •

TEETH $\frac{56}{56}$

strongly curved flippers •

long flippers black on both sides •

white underside •

pointed tips •

BEHAVIOR

Normally a boisterous swimmer and capable of speeds exceeding 12 knots. Known to ride bow waves and stern waves of fast boats and ships, swimming with long, low, leaps. From a distance, this undulating motion makes it look like a swimming penguin. Will swim alongside slow vessels. When swimming fast may travel very close to the surface without actually leaving the water, creating a great deal of spray when it rises to breathe. Has been observed spinning around on longitudinal axis when riding waves. May associate with other pelagic species, such as Fin Whales, Sei Whales, Southern Bottlenose Whales, Arnoux's Beaked Whales, Killer Whales, Long-finned Pilot Whales, and Southern Rightwhale Dolphins.

ID CHECKLIST

• black and white markings
• hourglass pattern on sides
• short black beak
• prominent fin
• stocky body
• black flippers, fin, and flukes
• frequently bow-rides
• undulating swimming motion
• usually in small groups

Group size 1–7 (1–40), 1 exceptional case of about 100 reported together	Fin position Center

Status Locally common	Population Unknown	Threats Unknown

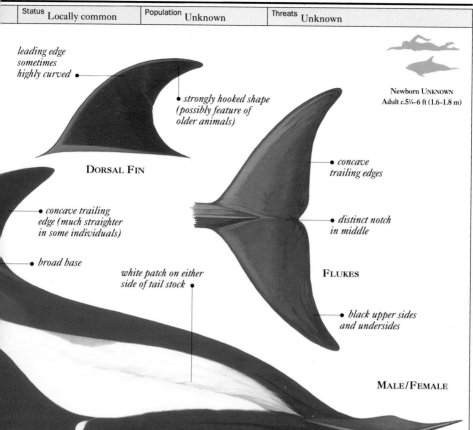

*leading edge
sometimes
highly curved* •

• *strongly hooked shape
(possibly feature of
older animals)*

Newborn UNKNOWN
Adult *c*.5¼–6 ft (1.6–1.8 m)

DORSAL FIN

• *concave
trailing edges*

• *concave trailing
edge (much straighter
in some individuals)*

• *distinct notch
in middle*

• *broad base*

*white patch on either
side of tail stock* •

FLUKES

• *black upper sides
and undersides*

MALE/FEMALE

• *conspicuous keel on tail
stock, especially underside*

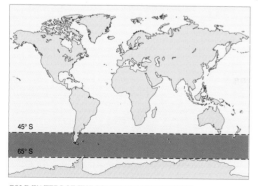

45° S

65° S

COLD WATERS OF THE SOUTHERN HEMISPHERE,
PREDOMINANTLY BETWEEN 45° S AND 65° S

WHERE TO LOOK

Distribution poorly known, though range
appears to be fairly extensive. Mostly
occurs in the South Atlantic and South
Pacific, and in cool currents associated
with the West-wind Drift. Occurs within
100 miles (160 km) of the ice edge in
some southern parts of range; northern
limits largely unknown, but probably
below 45° S. A single record as far north as
Valparaíso, Chile, seems to be exceptional.
Normally seen far out to sea; however, has
been observed in fairly shallow water near
the Antarctic Peninsula and off southern
South America. Range probably shifts
north and south with the seasons.

Birth wt Unknown	Adult wt 200–265 lb (90–120 kg)	Diet

Family DELPHINIDAE	Species *Lagenorhynchus obliquidens*	Habitat 〰〰 ▰〰

PACIFIC WHITE-SIDED DOLPHIN

The Pacific White-sided Dolphin is particularly lively. Large schools make so much disturbance in the water that their splashes can often be seen long before the animals themselves. The body pattern varies greatly between individuals and tends to be less distinctive in younger animals. It looks remarkably similar to the Dusky Dolphin (p.220), but there is no overlap in range. When swimming fast, the Pacific White-sided Dolphin may produce a spray of water known as a "rooster

tail," and so at a distance it could be mistaken for Dall's Porpoise (p.248). Confusion is most likely with the Common Dolphin (p.164), but the Pacific White-sided has a shorter beak and no hourglass pattern on its sides.
• **OTHER NAMES** Lag, Pacific Striped Dolphin, White-striped Dolphin, Hook-finned Porpoise.

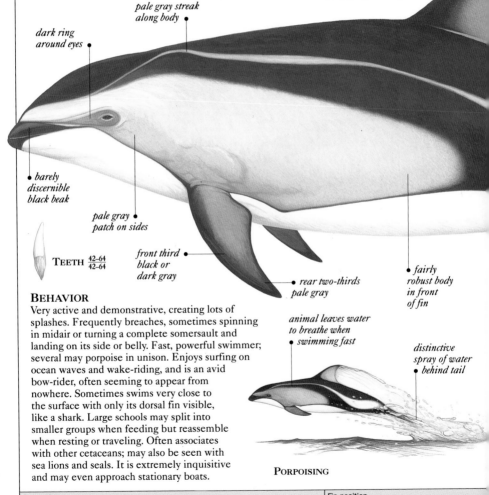

front third dark gray or black (variable) •

pale gray streak along body •

dark ring around eyes •

• *barely discernible black beak*

TEETH $\frac{42-64}{42-64}$

pale gray patch on sides •

front third black or dark gray •

• *rear two-thirds pale gray*

• *fairly robust body in front of fin*

BEHAVIOR

Very active and demonstrative, creating lots of splashes. Frequently breaches, sometimes spinning in midair or turning a complete somersault and landing on its side or belly. Fast, powerful swimmer; several may porpoise in unison. Enjoys surfing on ocean waves and wake-riding, and is an avid bow-rider, often seeming to appear from nowhere. Sometimes swims very close to the surface with only its dorsal fin visible, like a shark. Large schools may split into smaller groups when feeding but reassemble when resting or traveling. Often associates with other cetaceans; may also be seen with sea lions and seals. It is extremely inquisitive and may even approach stationary boats.

animal leaves water to breathe when • swimming fast

distinctive spray of water • behind tail

PORPOISING

Group size 10–100 (1–2,000), smaller groups may be found inshore	Fin position Center

Status Common	Population Unknown	Threats

ID CHECKLIST

- black or dark gray upper side
- white underside
- pale gray patch above flippers
- bicolored fin and flippers
- barely discernible black beak
- tall, falcate fin
- pale gray streak on sides
- usually in large groups
- acrobatic and demonstrative

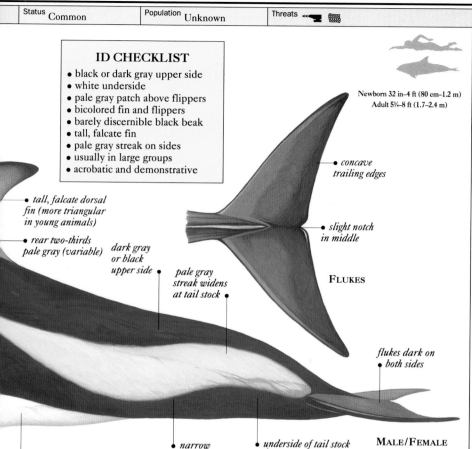

Newborn 32 in–4 ft (80 cm–1.2 m)
Adult 5¾–8 ft (1.7–2.4 m)

- concave trailing edges

- tall, falcate dorsal fin (more triangular in young animals)

- rear two-thirds pale gray (variable)

dark gray or black upper side

pale gray streak widens at tail stock

- slight notch in middle

FLUKES

flukes dark on both sides

- white underside

- narrow tail stock

- underside of tail stock black or dark gray

MALE/FEMALE

WHERE TO LOOK

Tends to remain south of colder waters influenced by arctic currents and stays north of the tropics. Although common in the Gulf of Alaska, and also around the southern Kamchatka Peninsula, it is absent from the Bering Sea. Mainly found offshore, as far as the edge of the continental shelf, but does come closer to shore where there is deep water, such as over submarine canyons. May be north–south or inshore–offshore movements with the seasons (may move inshore or south-ward in the winter), but some populations are probably resident year-round.

DEEP TEMPERATE WATERS OF THE NORTHERN NORTH PACIFIC, PREDOMINANTLY OFFSHORE

Birth wt c.35 lb (15 kg)	Adult wt 185–330 lb (85–150 kg)	Diet

Family DELPHINIDAE	Species *Lagenorhynchus obscurus*	Habitat

DUSKY DOLPHIN

One of the most acrobatic of all dolphins, the Dusky Dolphin is well known for its extraordinary high leaps and somersaults. It is highly gregarious and seems to welcome the company of other species as well as its own: it is often seen with seabirds and frequently associates with other cetaceans. Its own group sizes vary according to the time of year, with larger numbers living together during the summer; it tends to separate into small subgroups to feed, but these aggregate to socialize and rest. "Duskies" are remarkably similar to Pacific White-sided Dolphins (p.218) and, although there is no overlap in range, some experts have suggested that they may belong to the same species. Confusion is likely with Peale's Dolphin (p.214), although this has only a single stripe on each side and a dark face and throat. There are subtle variations in the pigmentation pattern of individual Dusky Dolphins and among different populations.

• **OTHER NAME** Fitzroy's Dolphin.

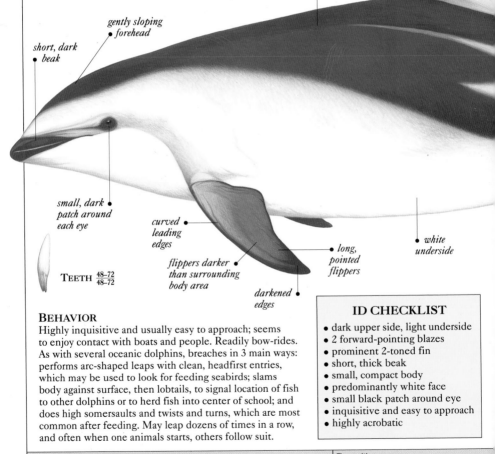

predominantly dark gray or blue-black upper side

gently sloping forehead

short, dark beak

small, dark patch around each eye

curved leading edges

flippers darker than surrounding body area

darkened edges

long, pointed flippers

white underside

TEETH $\frac{48-72}{48-72}$

BEHAVIOR
Highly inquisitive and usually easy to approach; seems to enjoy contact with boats and people. Readily bow-rides. As with several oceanic dolphins, breaches in 3 main ways: performs arc-shaped leaps with clean, headfirst entries, which may be used to look for feeding seabirds; slams body against surface, then lobtails, to signal location of fish to other dolphins or to herd fish into center of school; and does high somersaults and twists and turns, which are most common after feeding. May leap dozens of times in a row, and often when one animals starts, others follow suit.

ID CHECKLIST
• dark upper side, light underside
• 2 forward-pointing blazes
• prominent 2-toned fin
• short, thick beak
• small, compact body
• predominantly white face
• small black patch around eye
• inquisitive and easy to approach
• highly acrobatic

Group size 6–15 (6–500), largest gatherings in summer	Fin position Center

Status Locally common	Population Unknown	Threats

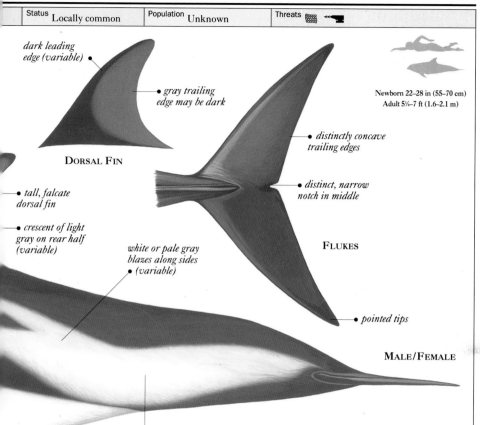

dark leading edge (variable)

gray trailing edge may be dark

DORSAL FIN

tall, falcate dorsal fin

crescent of light gray on rear half (variable)

white or pale gray blazes along sides (variable)

Newborn 22–28 in (55–70 cm)
Adult 5¼–7 ft (1.6–2.1 m)

distinctly concave trailing edges

distinct, narrow notch in middle

FLUKES

pointed tips

MALE/FEMALE

small, compact body

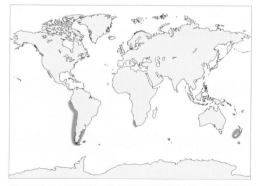

COASTAL TEMPERATE WATERS OF NEW ZEALAND, SOUTHERN AFRICA, AND SOUTH AMERICA

WHERE TO LOOK

Widespread in southern hemisphere, but distribution probably not continuous: the 3 main populations seem to be geographically isolated. Occurs around New Zealand, including the Chatham Islands, Auckland Islands, and Campbell Island; southern Africa; and South America, including the Falkland Islands. It is an infrequent visitor south of Valparaíso, Chile, and south of Peninsula Valdés, Argentina (though it is frequent at Valdés itself). Also occurs around Kerguelen Island, in the southern Indian Ocean. Some unconfirmed reports of individuals off the coast of Australia. Mainly coastal, or found over the continental shelf. Some inshore–offshore movements with seasons and times of day. Present in some areas year-round. Fairly abundant throughout range.

Birth wt 7–11 lb (3–5 kg)	Adult wt 110–200 lb (50–90 kg)	Diet

Family DELPHINIDAE	Species *Orcaella brevirostris*	Habitat 〰〰 〰〰

IRRAWADDY DOLPHIN

The Irrawaddy Dolphin is poorly known and easy to overlook. Best located by its loud blow, it is common in certain areas but probably not abundant over much of its range. Superficially similar in shape to the Beluga (p.92), it is sometimes placed in the same family, the Monodontidae; however, it shares characteristics with members of the dolphin family as well. It sometimes associates with Indo-Pacific Hump-backed Dolphins but is more likely to be confused with the Finless Porpoise (p.238), which also has a rounded, blunt head but is considerably smaller and does not have a dorsal fin. There is some overlap in range with the Dugong, a form of sea cow, which may also look similar from a distance. Its tropical river, estuary, and coastal habitat is very vulnerable to damming and other industrial development.
• **OTHER NAME** Snubfin Dolphin.

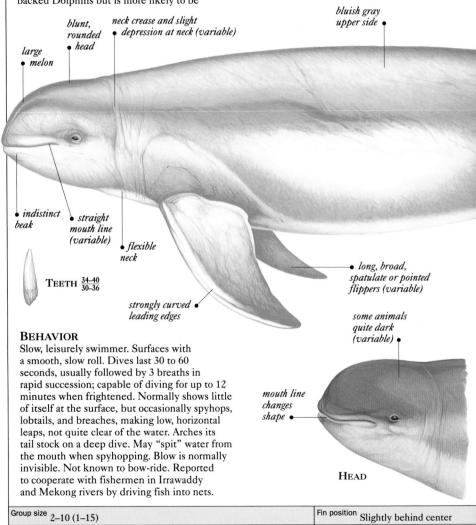

bluish gray upper side •

blunt, rounded • *head*

neck crease and slight • *depression at neck (variable)*

large • *melon*

• *indistinct beak*

• *straight mouth line (variable)*

• *flexible neck*

TEETH 34–40 / 30–36

strongly curved • *leading edges*

• *long, broad, spatulate or pointed flippers (variable)*

some animals quite dark (variable) •

BEHAVIOR
Slow, leisurely swimmer. Surfaces with a smooth, slow roll. Dives last 30 to 60 seconds, usually followed by 3 breaths in rapid succession; capable of diving for up to 12 minutes when frightened. Normally shows little of itself at the surface, but occasionally spyhops, lobtails, and breaches, making low, horizontal leaps, not quite clear of the water. Arches its tail stock on a deep dive. May "spit" water from the mouth when spyhopping. Blow is normally invisible. Not known to bow-ride. Reported to cooperate with fishermen in Irrawaddy and Mekong rivers by driving fish into nets.

mouth line changes shape •

HEAD

Group size 2–10 (1–15)	Fin position Slightly behind center

Status	Population	Threats
Locally common	Unknown	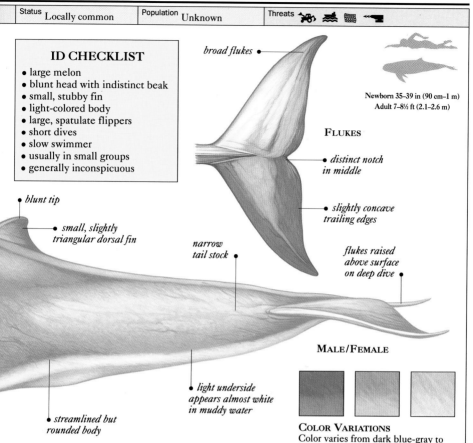

ID CHECKLIST

- large melon
- blunt head with indistinct beak
- small, stubby fin
- light-colored body
- large, spatulate flippers
- short dives
- slow swimmer
- usually in small groups
- generally inconspicuous

broad flukes

Newborn 35–39 in (90 cm–1 m)
Adult 7–8½ ft (2.1–2.6 m)

FLUKES

distinct notch in middle

blunt tip

small, slightly triangular dorsal fin

narrow tail stock

slightly concave trailing edges

flukes raised above surface on deep dive

MALE/FEMALE

light underside appears almost white in muddy water

streamlined but rounded body

COLOR VARIATIONS
Color varies from dark blue-gray to medium gray or pale blue. Underside is usually paler than the upper side.

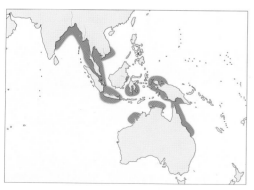

WARM COASTAL WATERS AND RIVERS FROM THE BAY OF BENGAL TO NORTHERN AUSTRALIA

WHERE TO LOOK

Mainly found in shallow coastal waters of the tropical Indo-Pacific but also in major river systems, especially: Brahmaputra and Ganges, India; Mekong, Vietnam, Laos, and Cambodia; Mahakam, Borneo; and Irrawaddy, Myanmar. Sometimes travels more than 805 miles (1,300 km) upstream; some individuals probably spend all their lives in fresh water. Along the coast it seems to prefer sheltered areas, such as turbid estuaries and mangrove swamps, and not yet found more than a few miles offshore. Probably occurs in Northern Australia and possibly the Philippines. Map shows probable range after linking sites where it is known to occur.

Birth wt	Adult wt	Diet
c.26 lb (12 kg)	200–330 lb (90–150 kg)	

RIVER DOLPHINS

THE RIVER DOLPHINS live in some of the largest, muddiest rivers in Asia and South America. They share many common features and have broadly similar habits, but may not be closely related. It is quite likely that they have simply adapted to their environments in similar ways, through convergent evolution. Despite their name, they are not all exclusively riverine animals, nor are they the only cetaceans living in rivers; Tucuxis, Finless Porpoises, and others regularly inhabit fresh water. The many threats to their survival include pollution, hunting, fishing, and dam construction.

CHARACTERISTICS

The geographical isolation of the river dolphins makes them easy to distinguish. There is also little room for confusion with the few oceanic dolphins that enter river systems, because the two groups are distinctly different both in appearance and in behavior. River dolphins are small animals – rarely over 8¼ ft (2.5 m) long – rather slow swimmers, and less inclined to leap than oceanic dolphins.

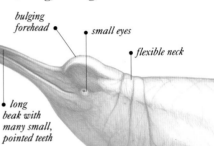

bulging forehead

small eyes

flexible neck

long beak with many small, pointed teeth

broad flippers

BEAK
River dolphins have long, narrow beaks that become proportionately longer with age. The Baiji (above) has a distinctive upturned beak.

BOTO
The most striking feature of most river dolphins is that they have tiny eyes and are almost blind. Good eyesight is virtually useless in turbid rivers, where visibility is very poor. Instead, they have a highly evolved echolocation system, which enables them to build up a "sound picture" of their surroundings.

BAIJI

back arches briefly before animal disappears

back and dorsal fin break surface

beak and forehead break surface first

flukes remain below surface when diving

DIVE SEQUENCE
River dolphins are undemonstrative compared to many oceanic dolphins. They usually show little of themselves at the surface but sometimes lift their long beaks into the air. Their dives usually last for no more than 40 seconds and are frequently much shorter.

ADAPTING TO HABITAT
The Boto is well adapted to life in flooded forests: its neck and flippers are sufficiently flexible for it to weave among the branches.

• *small or indistinct dorsal fin (except Franciscana)*

BLOWHOLE
River dolphins have a blowhole on the top of the head; it can be rounded, like that of the Baiji (below), crescent-shaped (Franciscana and Boto), or slitlike (Ganges and Indus River Dolphins). Some are easier to find by sound (they have a sneezelike blow) than by sight.

BAIJI (FROM ABOVE)

COLOR VARIATIONS
River dolphins show considerable color variation from one individual to the next. Their coloring often changes with age.

FRANCISCANA (p.234) *Inconspicuous dolphin; prefers coastal waters and does not inhabit rivers.*

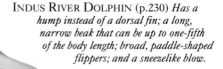

BAIJI (p.228) *Probably the most endangered cetacean; lives only in the Yangtze River, China.*

INDUS RIVER DOLPHIN (p.230) *Has a hump instead of a dorsal fin; a long, narrow beak that can be up to one-fifth of the body length; broad, paddle-shaped flippers; and a sneezelike blow.*

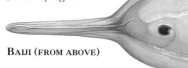

GANGES RIVER DOLPHIN (p.230) *Identical to the Indus River Dolphin but separated geographically. Officially, the two Asian river dolphins are different species, but some experts believe the evidence for splitting them is inconclusive.*

BOTO (p.226) *Largest of the group, with a long, fleshy dorsal ridge instead of a fin.*

SPECIES IDENTIFICATION

Family INIIDAE	Species *Inia geoffrensis*	Habitat

BOTO

This is the largest of the river dolphins and it is fairly easy to see. Three populations are recognized: in the Orinoco Basin, the Amazon Basin, and the upper Madeira River, South America. The populations have minor physical differences and are separated geographically, but may not be genetically distinct. The Boto often associates with the Tucuxi, the only other cetacean in the Amazon Basin, and sometimes shares feeding grounds with the Giant Otter. Its body color is highly variable according to age, water clarity, temperature, and location. The Boto population appears to be declining.

• **OTHER NAMES** Amazon River Dolphin, Pink Porpoise, Pink Dolphin.

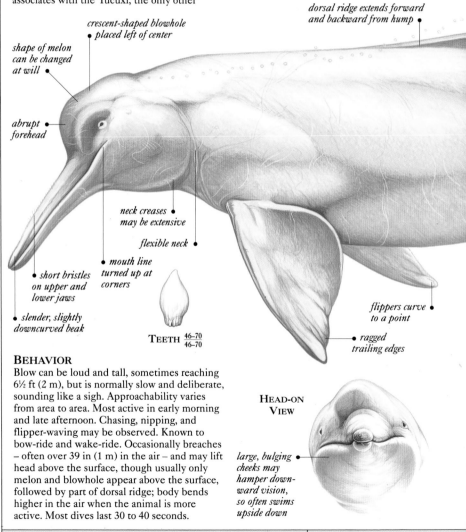

crescent-shaped blowhole
• placed left of center

shape of melon
can be changed
at will •

abrupt •
forehead

dorsal ridge extends forward
and backward from hump •

neck creases •
may be extensive

flexible neck •

• mouth line
turned up at
corners

• short bristles
on upper and
lower jaws

• slender, slightly
downcurved beak

TEETH $\frac{46-70}{46-70}$

flippers curve •
to a point

• ragged
trailing edges

BEHAVIOR
Blow can be loud and tall, sometimes reaching 6½ ft (2 m), but is normally slow and deliberate, sounding like a sigh. Approachability varies from area to area. Most active in early morning and late afternoon. Chasing, nipping, and flipper-waving may be observed. Known to bow-ride and wake-ride. Occasionally breaches – often over 39 in (1 m) in the air – and may lift head above the surface, though usually only melon and blowhole appear above the surface, followed by part of dorsal ridge; body bends higher in the air when the animal is more active. Most dives last 30 to 40 seconds.

HEAD-ON VIEW

large, bulging •
cheeks may
hamper down-
ward vision,
so often swims
upside down

Group size 1–2, up to 15 in dry season or at good feeding grounds (rare)	Hump position Slightly behind center

Status	Locally common	Population	Unknown	Threats

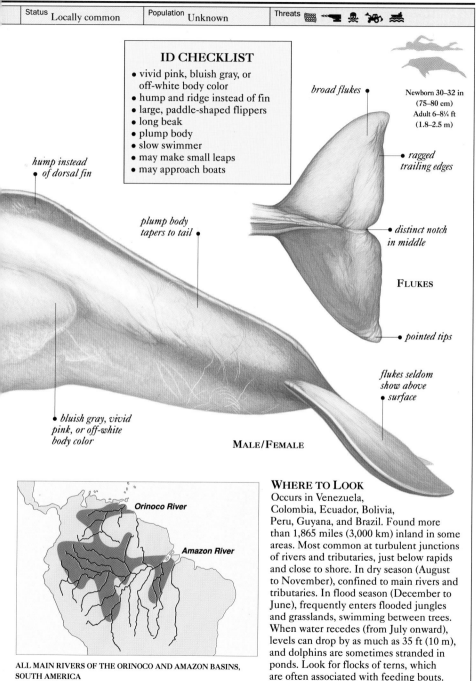

ID CHECKLIST

- vivid pink, bluish gray, or off-white body color
- hump and ridge instead of fin
- large, paddle-shaped flippers
- long beak
- plump body
- slow swimmer
- may make small leaps
- may approach boats

broad flukes •

Newborn 30–32 in (75–80 cm)
Adult 6–8¼ ft (1.8–2.5 m)

hump instead • of dorsal fin

• ragged trailing edges

plump body tapers to tail •

• distinct notch in middle

FLUKES

• pointed tips

flukes seldom show above • surface

• bluish gray, vivid pink, or off-white body color

MALE/FEMALE

WHERE TO LOOK

Occurs in Venezuela, Colombia, Ecuador, Bolivia, Peru, Guyana, and Brazil. Found more than 1,865 miles (3,000 km) inland in some areas. Most common at turbulent junctions of rivers and tributaries, just below rapids and close to shore. In dry season (August to November), confined to main rivers and tributaries. In flood season (December to June), frequently enters flooded jungles and grasslands, swimming between trees. When water recedes (from July onward), levels can drop by as much as 35 ft (10 m), and dolphins are sometimes stranded in ponds. Look for flocks of terns, which are often associated with feeding bouts.

Orinoco River

Amazon River

ALL MAIN RIVERS OF THE ORINOCO AND AMAZON BASINS, SOUTH AMERICA

Birth wt	c.15 lb (7 kg)	Adult wt	185–355 lb (85–160 kg)	Diet

Family PONTOPORIIDAE	Species *Lipotes vexillifer*	Habitat 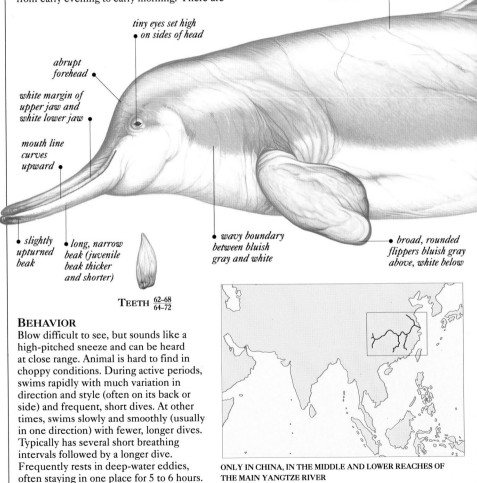

BAIJI

Little is known about the Baiji, as few specimens have been examined and it is a difficult animal to study in the wild (a single captive male, called Qi-Qi, has been a crucial source of information since 1980). It is easily frightened and usually impossible to approach by boat (it makes a long dive, changes direction underwater, swims underneath the boat, and surfaces far behind). Sometimes seen in the company of Finless Porpoises, it is most active from early evening to early morning. There are thought to be 40 to 45 separate groups altogether in the Yangtze River, China; but despite legal protection since 1949, the Baiji's population is still declining. It is probably the most endangered of all cetaceans.

• **OTHER NAMES** Yangtze River Dolphin, Beiji, Pei C'hi, Whitefin Dolphin, Whiteflag Dolphin, Chinese River Dolphin.

pale bluish gray upper side and sides •

tiny eyes set high on sides of head •

abrupt forehead •

white margin of upper jaw and white lower jaw •

mouth line curves upward •

• *slightly upturned beak*

• *long, narrow beak (juvenile beak thicker and shorter)*

• *wavy boundary between bluish gray and white*

• *broad, rounded flippers bluish gray above, white below*

TEETH $\frac{62-68}{64-72}$

BEHAVIOR
Blow difficult to see, but sounds like a high-pitched sneeze and can be heard at close range. Animal is hard to find in choppy conditions. During active periods, swims rapidly with much variation in direction and style (often on its back or side) and frequent, short dives. At other times, swims slowly and smoothly (usually in one direction) with fewer, longer dives. Typically has several short breathing intervals followed by a longer dive. Frequently rests in deep-water eddies, often staying in one place for 5 to 6 hours.

ONLY IN CHINA, IN THE MIDDLE AND LOWER REACHES OF THE MAIN YANGTZE RIVER

Group size 3–4 (1–6), more may gather at good feeding grounds	Fin position Slightly behind center

Status Endangered	Population c.150–200	Threats

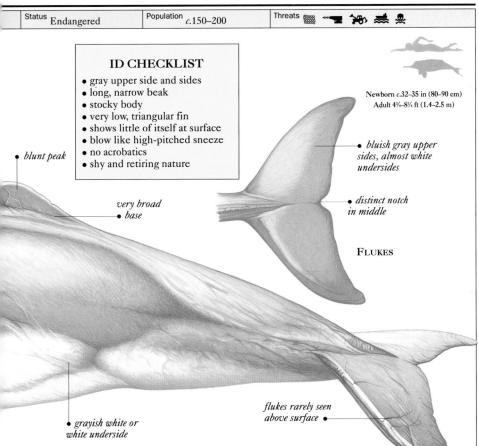

ID CHECKLIST

- gray upper side and sides
- long, narrow beak
- stocky body
- very low, triangular fin
- shows little of itself at surface
- blow like high-pitched sneeze
- no acrobatics
- shy and retiring nature

blunt peak

very broad base

Newborn c.32–35 in (80–90 cm)
Adult 4¾–8¼ ft (1.4–2.5 m)

bluish gray upper sides, almost white undersides

distinct notch in middle

FLUKES

grayish white or white underside

flukes rarely seen above surface

MALE/FEMALE

WHERE TO LOOK

Occurs along 1,055 miles (1,700 km) of the Yangtze River, China, though very rare above Zhicheng and below Nanjing. Most common from Luoshan to Xintankou, and from Anqing to Heishazhou. Found mainly where tributaries enter the river, especially immediately upstream or downstream of sandbanks and islets. When feeding, may come closer to shore, and often hunts over shallow sandbanks. Rarely found in areas without sandbanks. Once occurred in lakes Dongting and Poyang during spring floods, but water levels are no longer high enough. May migrate according to water levels. Semicaptive animals can be seen in nature reserves near Tongling and Shi-shou.

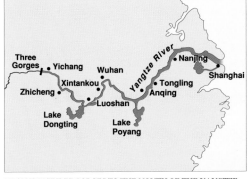

FROM THE THREE GORGES TO THE MOUTH OF THE YANGTZE RIVER, CHINA

Birth wt 6–11 lb (2.5–4.8 kg)	Adult wt 220–355 lb (100–160 kg)	Diet

Family PLATANISTIDAE	Species *Platanista minor* (Indus)	Habitat 〰️

INDUS AND GANGES RIVER DOLPHINS

For many years, these two river dolphins were thought to be the same species. Then, in the 1970s, differences in the structure of their skulls were discovered; recent research has also revealed blood protein differences, and they are now considered to be separate species. Although geographically separated, they look identical externally and have similar habits. Both species tend to live alone, or in pairs, but groups of up to 10 have been seen; the number living together may depend on their overall population size (reports from the 19th century, when they were much more common, mention "large schools"). These are the only cetaceans without a crystalline eye lens, effectively making them blind. They can probably detect the direction – and perhaps

intensity – of light, but they navigate and find food using a sophisticated echolocation system. Technical data for the Indus River Dolphin appears in the colored bands on these pages; for the Ganges River Dolphin, see pp.232–233.

• **OTHER NAMES** Indus Susu (Indus River Dolphin); Ganges Susu, Gangetic Dolphin (Ganges River Dolphin); Blind River Dolphin, Side-swimming Dolphin (both species).

longitudinal blowhole on left side

steeply sloping forehead

long, sharp front teeth visible even when mouth is closed

tiny eyes

mouth line curves upward

beak thickens slightly toward end

broad, paddle-shaped flippers showing "fingers"

undulating edges

TEETH $\frac{52-78}{52-70}$

BEHAVIOR

Both species swim and vocalize around the clock, with no obvious rest periods. They show more of themselves above the surface than other river dolphins and sometimes swim with their beaks sticking out of the water. They may breach when distressed, rising almost entirely out of the water and returning head first, usually accompanied by a loud lobtail. Females may lift calves above the surface on their backs. The usual interval between surfacings is 30 to 45 seconds, and animals often change direction immediately after disappearing underwater. Usually slow-moving but capable of bursts of speed.

beak longer than male's

FEMALE

Group size 1–2 (1–10)	Hump position Far behind center

Status Endangered	Population *c.*500	Threats ☠ ▦ ◄■ 🐙 ≈

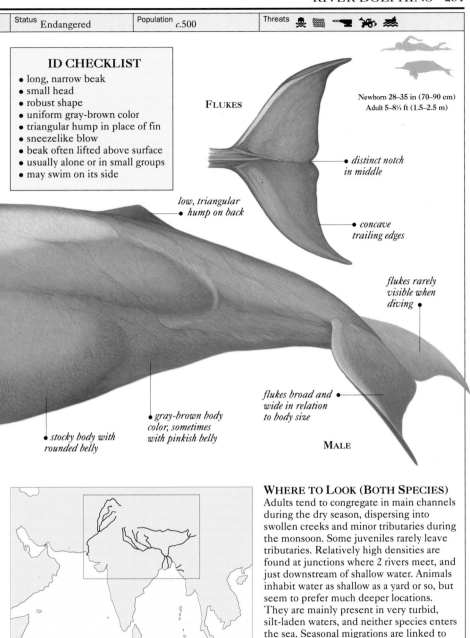

ID CHECKLIST

- long, narrow beak
- small head
- robust shape
- uniform gray-brown color
- triangular hump in place of fin
- sneezelike blow
- beak often lifted above surface
- usually alone or in small groups
- may swim on its side

FLUKES

Newborn 28–35 in (70–90 cm)
Adult 5–8¼ ft (1.5–2.5 m)

• *distinct notch in middle*

low, triangular • *hump on back*

• *concave trailing edges*

flukes rarely visible when diving •

flukes broad and • *wide in relation to body size*

• *gray-brown body color, sometimes with pinkish belly*

• *stocky body with rounded belly*

MALE

WHERE TO LOOK (BOTH SPECIES)

Adults tend to congregate in main channels during the dry season, dispersing into swollen creeks and minor tributaries during the monsoon. Some juveniles rarely leave tributaries. Relatively high densities are found at junctions where 2 rivers meet, and just downstream of shallow water. Animals inhabit water as shallow as a yard or so, but seem to prefer much deeper locations. They are mainly present in very turbid, silt-laden waters, and neither species enters the sea. Seasonal migrations are linked to the monsoon, which affects accessible areas. Some animals travel upstream as water levels rise, but barrages have possibly disrupted traditionally longer migrations.

INDUS, GANGES, BRAHMAPUTRA, AND MEGHNA RIVERS OF PAKISTAN, INDIA, BANGLADESH, NEPAL, AND BHUTAN

Birth wt 17 lb (7.5 kg)	Adult wt 155–200 lb (70–90 kg)	Diet ⟝ 🐟

Family PLATANISTIDAE	Species *Platanista gangetica* (Ganges)	Habitat

SIDE-SWIMMING

Both species sometimes swim on one side, especially in shallow water. Animals usually lean to the right and cruise near the bottom of the river; the tail is held slightly higher than the head (which nods continuously), and one flipper is often trailed in the mud to search for food. According to some reports, more side-swimming takes place in the evening, and it reaches a peak at night. They frequently swim in circles underwater, usually counterclockwise.

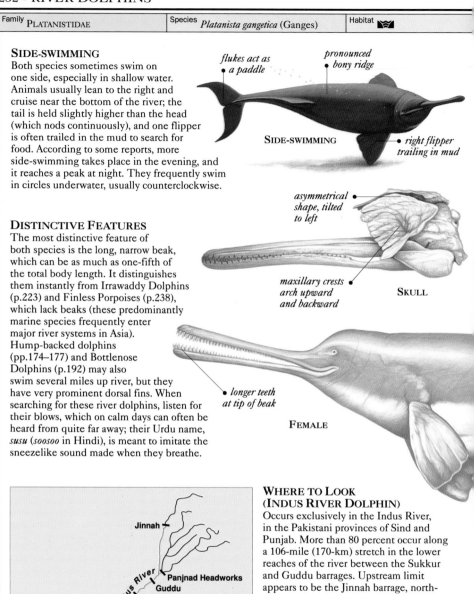

flukes act as a paddle

pronounced bony ridge

SIDE-SWIMMING

right flipper trailing in mud

asymmetrical shape, tilted to left

maxillary crests arch upward and backward

SKULL

longer teeth at tip of beak

FEMALE

DISTINCTIVE FEATURES

The most distinctive feature of both species is the long, narrow beak, which can be as much as one-fifth of the total body length. It distinguishes them instantly from Irrawaddy Dolphins (p.223) and Finless Porpoises (p.238), which lack beaks (these predominantly marine species frequently enter major river systems in Asia). Hump-backed dolphins (pp.174–177) and Bottlenose Dolphins (p.192) may also swim several miles up river, but they have very prominent dorsal fins. When searching for these river dolphins, listen for their blows, which on calm days can often be heard from quite far away; their Urdu name, *susu* (*soosoo* in Hindi), is meant to imitate the sneezelike sound made when they breathe.

WHERE TO LOOK (INDUS RIVER DOLPHIN)

Occurs exclusively in the Indus River, in the Pakistani provinces of Sind and Punjab. More than 80 percent occur along a 106-mile (170-km) stretch in the lower reaches of the river between the Sukkur and Guddu barrages. Upstream limit appears to be the Jinnah barrage, northwestern Punjab; downstream limit is likely to be the Kotri barrage, Sind. Also inhabits the Chenab River, below Panjnad headworks. Since the 1930s, many barrages built for irrigation and hydroelectric power generation have severely affected its movement and distribution, splitting the population into isolated pockets.

Jinnah

Indus River Panjnad Headworks
Guddu
Sukkur

Kotri

Arabian Sea

INDUS RIVER IN PAKISTAN FROM KOTRI, SIND, TO JINNAH, NORTHWESTERN PUNJAB

Group size 1–2 (1–10)	Hump position Far behind center

Status Endangered	Population 4,000–6,000	Threats ☠ 🔲 ◀🟥 🐾 🌊

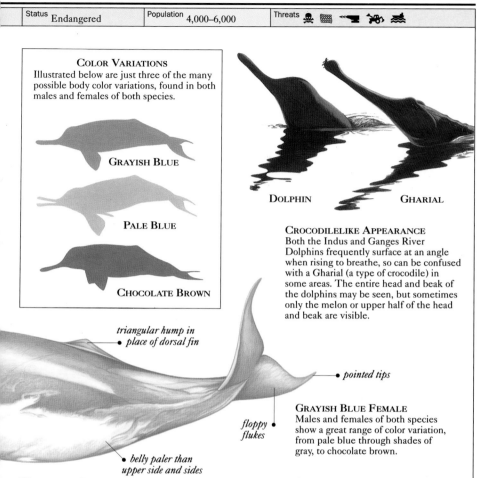

COLOR VARIATIONS
Illustrated below are just three of the many possible body color variations, found in both males and females of both species.

GRAYISH BLUE

PALE BLUE

CHOCOLATE BROWN

DOLPHIN   GHARIAL

CROCODILELIKE APPEARANCE
Both the Indus and Ganges River Dolphins frequently surface at an angle when rising to breathe, so can be confused with a Gharial (a type of crocodile) in some areas. The entire head and beak of the dolphins may be seen, but sometimes only the melon or upper half of the head and beak are visible.

triangular hump in place of dorsal fin

pointed tips

floppy flukes

GRAYISH BLUE FEMALE
Males and females of both species show a great range of color variation, from pale blue through shades of gray, to chocolate brown.

belly paler than upper side and sides

WHERE TO LOOK (GANGES RIVER DOLPHIN)
More widely distributed than the Indus River Dolphin, occurring in the Ganges, Meghna, and Brahmaputra river systems of western India, Nepal, Bhutan, and Bangladesh, and the Karnaphuli River, Bangladesh. It may also inhabit the Sangu River, Bangladesh, and Chinese head-waters of the Brahmaputra. Discontinuous distribution from the foothills of the Himalayas to the limits of the tidal zone. Upstream limits appear to be Sunamganj (Meghna), Lohit River (Brahmaputra), and Manao (Ganges). Does not go down-stream beyond tidal limits. The Farakka barrage has split the population in two.

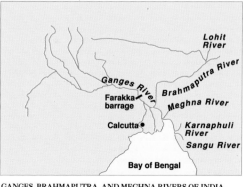

GANGES, BRAHMAPUTRA, AND MEGHNA RIVERS OF INDIA, BANGLADESH, NEPAL, AND BHUTAN

Birth wt 17 lb (7.5 kg)	Adult wt 155–200 lb (70–90 kg)	Diet 🐟 🦐

Family PONTOPORIIDAE	Species *Pontoporia blainvillei*	Habitat 〰〰

FRANCISCANA

There are few records of the Franciscana in the wild. It is an inconspicuous and undemonstrative animal, easy to overlook in anything but the calmest conditions. Although it is closely related to the river dolphins, it lives in the sea and prefers very shallow coastal waters. One of the smallest of all cetaceans, its most unusual feature is the beak, which is the longest (relative to body size) of any dolphin, though the beak of the juvenile is considerably shorter than that of

the adult. Its body color may lighten during winter and with age; some older animals are predominantly white. The limited number of observations suggest that it is usually a solitary animal, although groups of up to 5 dolphins have been reported. Entanglement in fishing nets is a major cause of death and has almost certainly depleted the species.
• **OTHER NAME** La Plata Dolphin.

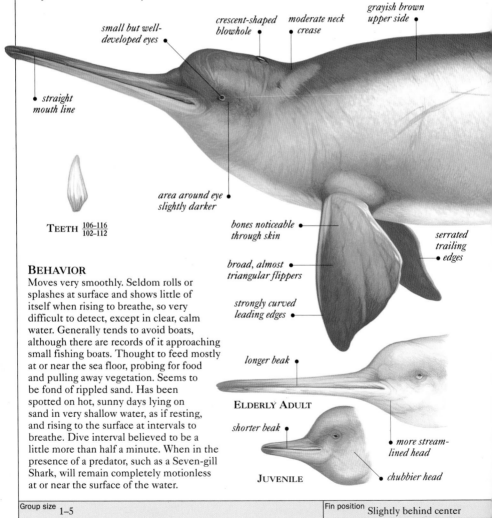

small but well-developed eyes •

crescent-shaped blowhole •

moderate neck • crease

grayish brown upper side •

• straight mouth line

area around eye • slightly darker

TEETH 106–116 / 102–112

bones noticeable • through skin

serrated trailing • edges

broad, almost • triangular flippers

strongly curved leading edges •

longer beak •

ELDERLY ADULT

shorter beak •

• more stream-lined head

JUVENILE

• chubbier head

BEHAVIOR

Moves very smoothly. Seldom rolls or splashes at surface and shows little of itself when rising to breathe, so very difficult to detect, except in clear, calm water. Generally tends to avoid boats, although there are records of it approaching small fishing boats. Thought to feed mostly at or near the sea floor, probing for food and pulling away vegetation. Seems to be fond of rippled sand. Has been spotted on hot, sunny days lying on sand in very shallow water, as if resting, and rising to the surface at intervals to breathe. Dive interval believed to be a little more than half a minute. When in the presence of a predator, such as a Seven-gill Shark, will remain completely motionless at or near the surface of the water.

Group size 1–5	Fin position Slightly behind center

Status Locally common	Population Unknown	Threats 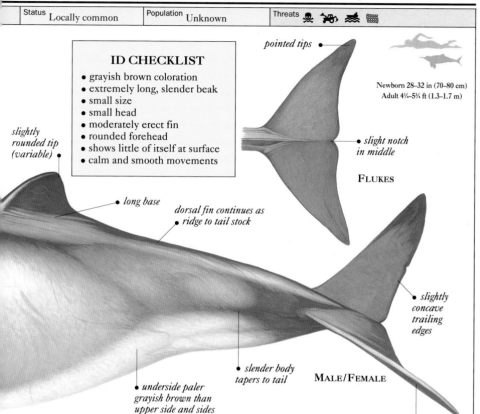

ID CHECKLIST

- grayish brown coloration
- extremely long, slender beak
- small size
- small head
- moderately erect fin
- rounded forehead
- shows little of itself at surface
- calm and smooth movements

pointed tips

Newborn 28–32 in (70–80 cm)
Adult 4¼–5¼ ft (1.3–1.7 m)

slight notch in middle

FLUKES

slightly rounded tip (variable)

long base

dorsal fin continues as ridge to tail stock

slightly concave trailing edges

slender body tapers to tail

MALE/FEMALE

underside paler grayish brown than upper side and sides

extremely broad flukes; width up to one-third of body length

WHERE TO LOOK

This is the only member of the river dolphin family living in the sea. However, it prefers shallow coastal waters. Most sightings are close to land, usually where depth is no more than about 30 ft (9 m). Known range extends from the Doce River, near Regencia, Brazil, south through Uruguay to Bahía Blanca, Argentina; may occur as far south as the northern coast of Golfo San Matias, Argentina. Once occurred as far south as Peninsula Valdés, Argentina, but is rarely sighted there nowadays. Most common on the Uruguayan side of the La Plata estuary. Although common in the La Plata estuary, does not inhabit rivers and has never been recorded further upstream than Buenos Aires, Argentina. Rarely seen in winter, suggesting some kind of seasonal movement.

TEMPERATE COASTAL WATERS OF EASTERN SOUTH AMERICA

Birth wt 16–19 lb (7.3–8.5 kg)	Adult wt 65–115 lb (30–53 kg)	Diet

PORPOISES

PATIENCE, perseverance, and a certain amount of luck are prerequisites for watching most porpoises. Consequently, they are all too often overlooked by many whale watchers. They are threatened by a variety of human activities and, sadly, many of their populations are declining; they are especially susceptible to entangle-ment and drowning in fishing nets. They live mainly along the coast, but also in some rivers and in the open sea. "Porpoise" tends to be a general term used, particularly in North America, to mean any small dolphin; however, it is in fact the name for the 6 species in this family. Among them are some of the smallest cetaceans in the world.

CHARACTERISTICS
Preferring to keep to themselves, porpoises are typically shy creatures and rarely perform the acrobatic feats of dolphins. They tend to live alone or in small groups and, with the exception of Dall's Porpoise and some Finless Porpoises, are usually wary of boats; as a result, most porpoises are poorly known. A brief glimpse of the dorsal fin and a small portion of the back is all that is normally seen. Typically smaller and more robust than dolphins, they rarely grow beyond 6½ ft (2 m) and, at sea, show so little of themselves at the surface that they appear even smaller. In many places, porpoises can be distinguished from one another by geography: as a family they are widely distributed, but as there is little overlap in range, the species can often be identified by a process of elimination.

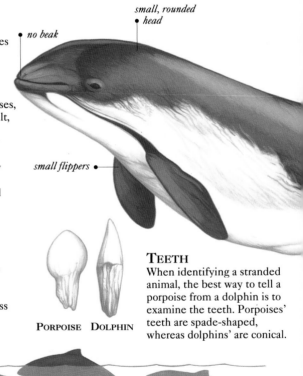

small, rounded head

no beak

small flippers

PORPOISE   DOLPHIN

TEETH
When identifying a stranded animal, the best way to tell a porpoise from a dolphin is to examine the teeth. Porpoises' teeth are spade-shaped, whereas dolphins' are conical.

DALL'S PORPOISE

flukes usually stay below surface when diving

dorsal fin and small part of back are often all that can be seen at surface

most porpoises surface in a slow, smooth motion

animal disappears with little disturbance of water

DIVE SEQUENCE
When traveling slowly, most porpoises surface to breathe in a smooth, forward-rolling motion. When diving, the flukes usually stay below the surface. The blow is rarely seen but may be heard.

VAQUITA (p.244) *Only porpoise in the northern end of the Gulf of California; occurs nowhere else.*

DALL'S PORPOISE
Dall's Porpoise is the odd member of the family. It prefers deep water and has horny protuberances on its gums, between its normal teeth. It produces this distinctive spray of water when it surfaces to breathe.

FINLESS PORPOISE (p.238) *Only porpoise without a dorsal fin; very streamlined body.*

well-defined dorsal fin
(except Finless Porpoise)

HARBOR PORPOISE
The Harbor Porpoise shows a number of physical features common to most porpoises, but it is difficult to see many of these under field conditions.

HARBOR PORPOISE (p.242) *Has a sharp, sneezelike blow and rather nondescript coloring.*

BURMEISTER'S PORPOISE (p.246) *Unusual, backward-leaning dorsal fin toward the tail end.*

notched flukes

TUBERCLES
Circular bumps, known as tubercles, occur along the leading edges of the dorsal fins and flippers of the Harbor and Burmeister's Porpoises. The Finless Porpoise has tubercles on its back, in place of a dorsal fin; these vary from a narrow row to a band 3–4 in (7–10 cm) wide at the front and tapering toward the tail. Tubercles are rarely visible under field conditions.

FIN WITH TUBERCLES

SPECTACLED PORPOISE (p.240) *Striking black and white markings and black "eye patch."*

DALL'S PORPOISE (p.248) *Fast-swimming, sociable porpoise that loves to bow-ride.*

Family PHOCOENIDAE	Species *Neophocaena phocaenoides*	Habitat

FINLESS PORPOISE

The Finless Porpoise is one of the smallest cetaceans. It is generally shy and difficult to approach almost everywhere except in the Yangtze River, China, where it appears to be used to heavy boat traffic. It is quite an active animal, usually swimming just beneath the surface with sudden, darting movements. Never found far from land, it is capable of surviving in extremely shallow tidal water. It is the only porpoise in the region, and the only one to have a bulbous melon. The Irrawaddy Dolphin (p.222) is similar in shape but has a stubby fin. It also resembles a small Beluga (p.92), but their ranges do not overlap. Its other names, Black Porpoise and Black Finless Porpoise, are misnomers: the body color only becomes black after death, and earlier descriptions were from dead animals rather than live ones. The body color may, however, darken slightly with age. The species was originally described from a specimen collected in South Africa, but this was probably mistakenly labeled.

• **OTHER NAMES** Black Porpoise, Black Finless Porpoise, Jiangzhu.

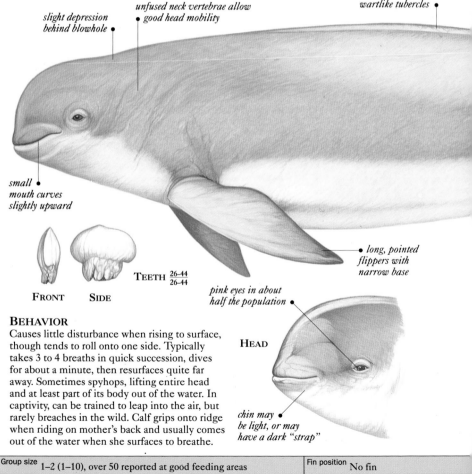

slight depression behind blowhole •

unfused neck vertebrae allow • good head mobility

ridge covered in circular, wartlike tubercles •

small • mouth curves slightly upward

TEETH $\frac{26-44}{26-44}$

FRONT   SIDE

pink eyes in about half the population •

long, pointed flippers with narrow base

HEAD

chin may • be light, or may have a dark "strap"

BEHAVIOR

Causes little disturbance when rising to surface, though tends to roll onto one side. Typically takes 3 to 4 breaths in quick succession, dives for about a minute, then resurfaces quite far away. Sometimes spyhops, lifting entire head and at least part of its body out of the water. In captivity, can be trained to leap into the air, but rarely breaches in the wild. Calf grips onto ridge when riding on mother's back and usually comes out of the water when she surfaces to breathe.

Group size 1–2 (1–10), over 50 reported at good feeding areas	Fin position No fin

Status	Population	Threats
Locally common	Unknown	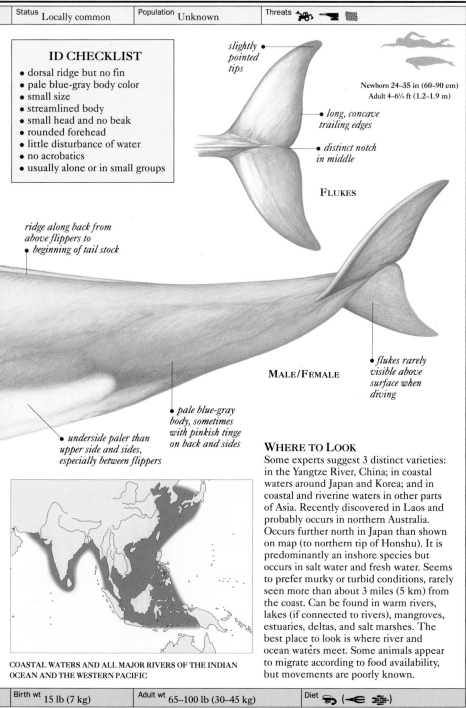

ID CHECKLIST

- dorsal ridge but no fin
- pale blue-gray body color
- small size
- streamlined body
- small head and no beak
- rounded forehead
- little disturbance of water
- no acrobatics
- usually alone or in small groups

slightly pointed tips

Newborn 24–35 in (60–90 cm)
Adult 4–6¼ ft (1.2–1.9 m)

- *long, concave trailing edges*
- *distinct notch in middle*

FLUKES

ridge along back from above flippers to beginning of tail stock

MALE/FEMALE

- *flukes rarely visible above surface when diving*

- *underside paler than upper side and sides, especially between flippers*

- *pale blue-gray body, sometimes with pinkish tinge on back and sides*

COASTAL WATERS AND ALL MAJOR RIVERS OF THE INDIAN OCEAN AND THE WESTERN PACIFIC

WHERE TO LOOK

Some experts suggest 3 distinct varieties: in the Yangtze River, China; in coastal waters around Japan and Korea; and in coastal and riverine waters in other parts of Asia. Recently discovered in Laos and probably occurs in northern Australia. Occurs further north in Japan than shown on map (to northern tip of Honshu). It is predominantly an inshore species but occurs in salt water and fresh water. Seems to prefer murky or turbid conditions, rarely seen more than about 3 miles (5 km) from the coast. Can be found in warm rivers, lakes (if connected to rivers), mangroves, estuaries, deltas, and salt marshes. The best place to look is where river and ocean waters meet. Some animals appear to migrate according to food availability, but movements are poorly known.

Birth wt	Adult wt	Diet
15 lb (7 kg)	65–100 lb (30–45 kg)	

Family PHOCOENIDAE	Species *Australophocaena dioptrica*	Habitat

SPECTACLED PORPOISE

The Spectacled Porpoise is poorly known. By the mid-1970s, only 10 specimens had been discovered. Since then, intensive searches have revealed more than 100 others, mostly from strandings along the wild Atlantic beaches of Tierra del Fuego, in southern South America; most reports are of animals that were already dead and often in an advanced state of decomposition. The Spectacled Porpoise is rarely sighted at sea but is believed to be more common than the lack of information suggests.

The striking black and white markings are quite distinctive, and it is one of the largest members of the porpoise family. There is a marked difference between the sexes, the dorsal fin of the male being much larger and more rounded than that of the female.
• **OTHER NAME** Formerly *Phocoena dioptrica*.

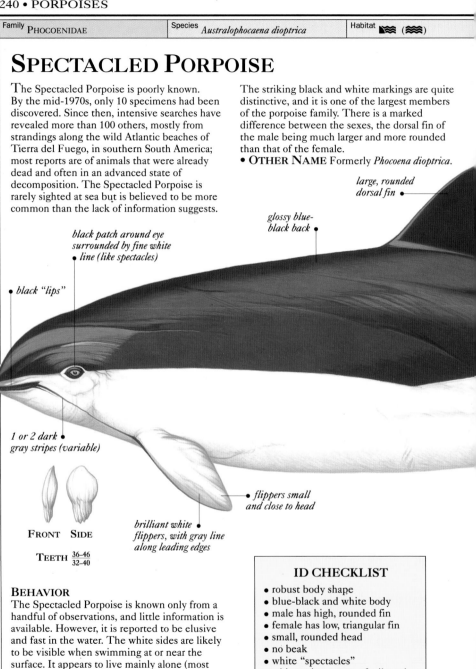

large, rounded
dorsal fin •

glossy blue-
black back •

black patch around eye
surrounded by fine white
• line (like spectacles)

• black "lips"

1 or 2 dark •
gray stripes (variable)

• flippers small
and close to head

FRONT   SIDE

brilliant white •
flippers, with gray line
along leading edges

TEETH 36–46
      32–40

BEHAVIOR
The Spectacled Porpoise is known only from a handful of observations, and little information is available. However, it is reported to be elusive and fast in the water. The white sides are likely to be visible when swimming at or near the surface. It appears to live mainly alone (most of the strandings and sightings are of solitary animals) but may also live in small groups.

ID CHECKLIST
• robust body shape
• blue-black and white body
• male has high, rounded fin
• female has low, triangular fin
• small, rounded head
• no beak
• white "spectacles"
• white stripe on top of tail stock

Group size 1–2 (1–10)	Fin position Slightly behind centre

Status Locally common	Population Unknown	Threats

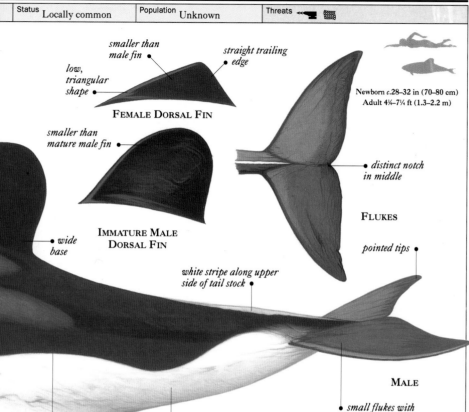

smaller than male fin •

straight trailing edge •

low, triangular shape •

FEMALE DORSAL FIN

Newborn *c.*28–32 in (70–80 cm)
Adult 4¼–7¼ ft (1.3–2.2 m)

smaller than mature male fin •

• *distinct notch in middle*

FLUKES

• *wide base*

IMMATURE MALE DORSAL FIN

pointed tips •

white stripe along upper side of tail stock •

MALE

• *small flukes with blue-black upper sides and white or pale gray undersides*

• *brilliant white underside extends to midway up flanks; white area may expand with age*

• *sharp demarcation between black and white*

SOUTHERN ATLANTIC COAST OF SOUTH AMERICA AND
CERTAIN OFFSHORE ISLANDS

WHERE TO LOOK
Most records of sightings and strandings
are from along the southern Atlantic
coast of South America. However,
distribution of this species is puzzling
because there are records from widely
separate locations; some of these may
involve strays or cases of mistaken
identity. Records from offshore islands
(mostly of dead animals and skulls), hint
at a circumpolar distribution and suggest
that the range may also include large
areas of open sea. It is not known whether
these represent isolated populations or
whether they mix with mainland coastal
animals by migrating across the open sea.

Birth wt Unknown	Adult wt 130–185 lb (60–84 kg)	Diet

Family PHOCOENIDAE	Species *Phocoena phocoena*	Habitat ≋≋

HARBOR PORPOISE

The Harbor Porpoise is difficult to observe. It shows little of itself at the surface, so a brief glimpse is the most common sighting. On calm days it may be possible to approach a basking animal, but it is generally wary of boats and rarely bow-rides. It can sometimes be detected by the blow, which, although rarely seen, makes a sharp, puffing sound rather like a sneeze. When it rises to breathe, the lasting impression is of a slow, forward-rolling motion, as if the dorsal fin is mounted on a revolving wheel lifted briefly above the surface and then withdrawn.

When feeding, or swimming fast, its whole body leaves the water so quickly that it is almost impossible to see what is happening. The dorsal fin is small but can appear large when scaled to the relatively small portion of visible back.

• **OTHER NAMES** Common Porpoise, Puffing Pig.

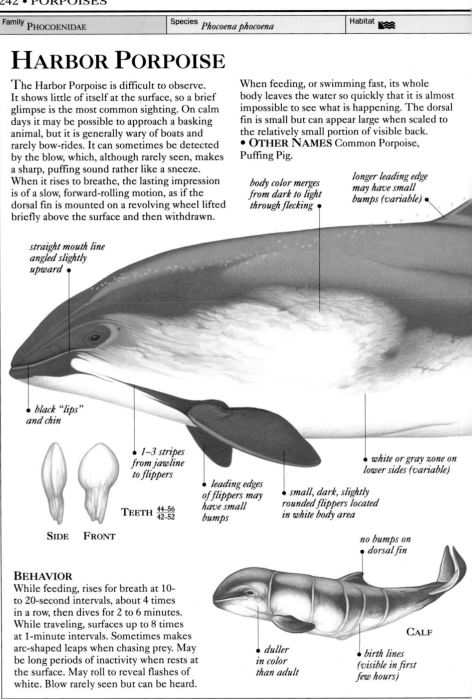

body color merges from dark to light through flecking

longer leading edge may have small bumps (variable)

straight mouth line angled slightly upward

black "lips" and chin

1–3 stripes from jawline to flippers

leading edges of flippers may have small bumps

TEETH $\frac{44-56}{42-52}$

SIDE   FRONT

small, dark, slightly rounded flippers located in white body area

white or gray zone on lower sides (variable)

no bumps on dorsal fin

CALF

BEHAVIOR
While feeding, rises for breath at 10- to 20-second intervals, about 4 times in a row, then dives for 2 to 6 minutes. While traveling, surfaces up to 8 times at 1-minute intervals. Sometimes makes arc-shaped leaps when chasing prey. May be long periods of inactivity when rests at the surface. May roll to reveal flashes of white. Blow rarely seen but can be heard.

duller in color than adult

birth lines (visible in first few hours)

Group size 2–5 (1–12), several hundred at good feeding grounds (rare)	Fin position Slightly behind center

Status Locally common	Population Unknown	Threats ☠ 🚜 🌊 🗔 ◄🔻

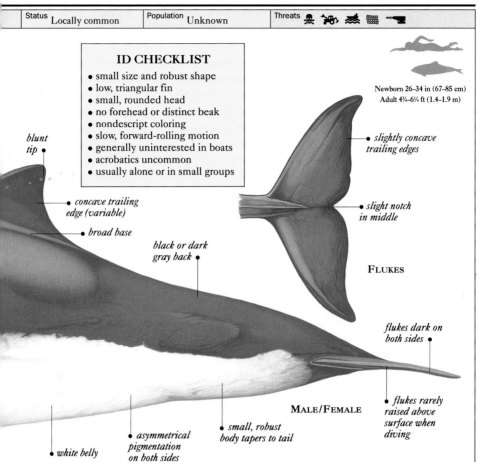

ID CHECKLIST

- small size and robust shape
- low, triangular fin
- small, rounded head
- no forehead or distinct beak
- nondescript coloring
- slow, forward-rolling motion
- generally uninterested in boats
- acrobatics uncommon
- usually alone or in small groups

Newborn 26–34 in (67–85 cm)
Adult 4¾–6¼ ft (1.4–1.9 m)

• *slightly concave trailing edges*

• *slight notch in middle*

FLUKES

blunt tip •

• *concave trailing edge (variable)*

• *broad base*

black or dark gray back •

flukes dark on both sides •

MALE/FEMALE

• *flukes rarely raised above surface when diving*

• *small, robust body tapers to tail*

• *white belly* • *asymmetrical pigmentation on both sides*

WHERE TO LOOK

Found in coastal waters, with most sightings within 6 miles (10 km) of land. Likes cool water and frequents relatively shallow bays, estuaries, and tidal channels under about 655 ft (200 m) in depth. Will swim a considerable distance upriver. Some seasonal movements (related to food availability) occur: mostly inshore in summer and offshore in winter, but sometimes north in summer and south in winter. In some areas, populations are present year-round. Black Sea, North Atlantic, and North Pacific populations are semi-isolated and have been proposed as separate subspecies. Some populations have become rarer in the past few decades.

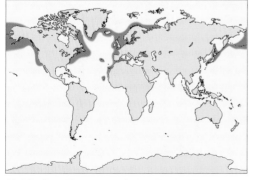

COLD TEMPERATE AND SUBARCTIC WATERS OF THE
NORTHERN HEMISPHERE

Birth wt 11 lb (5 kg)	Adult wt 125–145 lb (55–65 kg)	Diet 🐟 (◄ 🦑)

Family PHOCOENIDAE	Species *Phocoena sinus*	Habitat 〰〰

VAQUITA

Possibly the smallest of all cetaceans, the Vaquita is not easily confused with any other species within its limited range, but is rarely observed in the wild. Its complex but subdued gray patterning can appear olive or tawny brown under certain light conditions: many observers describe the overall impression as "dark." It is often called the Cochito, but this name can cause some confusion, as it is used by local fishermen to describe any small cetacean. The species is on the brink of extinction.
• OTHER NAMES Cochito, Gulf of California Porpoise.

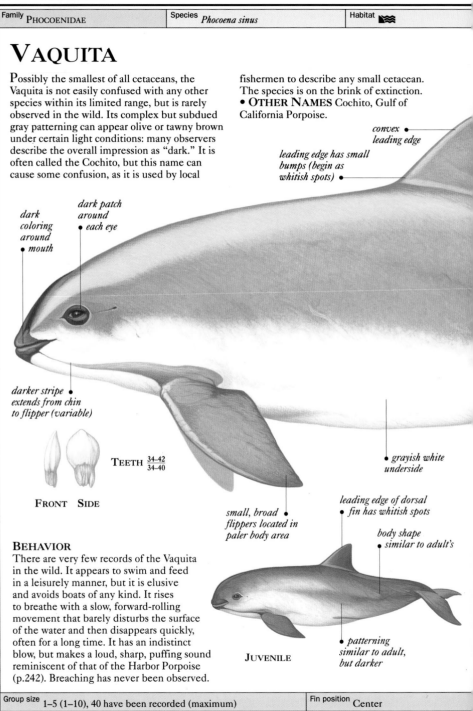

convex • leading edge

leading edge has small bumps (begin as whitish spots) •

dark coloring around • mouth

dark patch around • each eye

darker stripe • extends from chin to flipper (variable)

TEETH $\frac{34-42}{34-40}$

FRONT   SIDE

• grayish white underside

small, broad • flippers located in paler body area

leading edge of dorsal • fin has whitish spots

body shape • similar to adult's

BEHAVIOR
There are very few records of the Vaquita in the wild. It appears to swim and feed in a leisurely manner, but it is elusive and avoids boats of any kind. It rises to breathe with a slow, forward-rolling movement that barely disturbs the surface of the water and then disappears quickly, often for a long time. It has an indistinct blow, but makes a loud, sharp, puffing sound reminiscent of that of the Harbor Porpoise (p.242). Breaching has never been observed.

JUVENILE

• patterning similar to adult, but darker

Group size 1–5 (1–10), 40 have been recorded (maximum)	Fin position Center

Status Endangered	Population 100–500	Threats

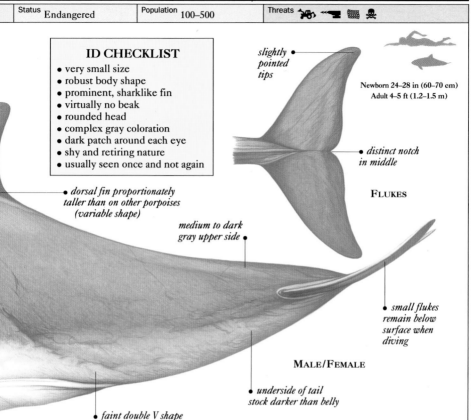

ID CHECKLIST
- very small size
- robust body shape
- prominent, sharklike fin
- virtually no beak
- rounded head
- complex gray coloration
- dark patch around each eye
- shy and retiring nature
- usually seen once and not again

slightly pointed tips

Newborn 24–28 in (60–70 cm)
Adult 4–5 ft (1.2–1.5 m)

• distinct notch in middle

FLUKES

• dorsal fin proportionately taller than on other porpoises (variable shape)

medium to dark gray upper side •

• small flukes remain below surface when diving

MALE/FEMALE

• underside of tail stock darker than belly

• faint double V shape pointing toward tail

WHERE TO LOOK

The Vaquita has the most limited distribution of any marine cetacean, occurring only in the extreme northern end of the Gulf of California (Sea of Cortez), western Mexico. It is most commonly found around the Colorado River delta. There may be slight seasonal movements north (in winter) and south (in summer), but there is little supporting data. The former range may have included an area farther south along the Mexican mainland. It lives in shallow, murky lagoons along the shoreline and is rarely seen in water much deeper than 95 ft (30 m); indeed, it can survive in lagoons so shallow that its back protrudes above the surface.

EXTREME NORTHERN END OF THE GULF OF CALIFORNIA
(SEA OF CORTEZ), MEXICO

Birth wt Unknown	Adult wt c.65–120 lb (30–55 kg)	Diet

Family PHOCOENIDAE	Species *Phocoena spinipinnis*	Habitat 〰️

BURMEISTER'S PORPOISE

Burmeister's Porpoise may be one of the most abundant small cetaceans living around the coasts of southern South America, but it is shy and easy to overlook, and so is poorly known. It is best distinguished from all other species by its small size and backward-pointing dorsal fin, lying in the rear third of the body. It overlaps in range with several small cetaceans but, on the Pacific coast at least, it is most likely to be confused with the Black Dolphin (p.200). However, the Black Dolphin's rounded dorsal fin, positioned only slightly behind the center

of its back, should be distinctive. Animals living along the Atlantic coast may be larger than those along the Pacific coast although, again, there is little information. They all turn completely black a few minutes after death.
• **OTHER NAME** Black Porpoise.

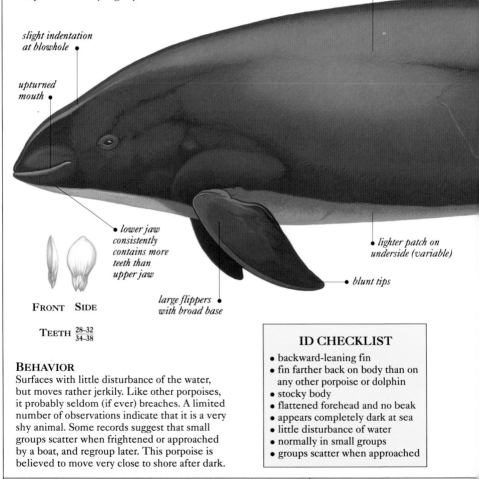

deep gray or black coloration may appear brown in certain lights

slight indentation at blowhole

upturned mouth

lower jaw consistently contains more teeth than upper jaw

FRONT   SIDE

large flippers with broad base

lighter patch on underside (variable)

blunt tips

TEETH $\frac{28-32}{34-38}$

BEHAVIOR
Surfaces with little disturbance of the water, but moves rather jerkily. Like other porpoises, it probably seldom (if ever) breaches. A limited number of observations indicate that it is a very shy animal. Some records suggest that small groups scatter when frightened or approached by a boat, and regroup later. This porpoise is believed to move very close to shore after dark.

ID CHECKLIST
- backward-leaning fin
- fin farther back on body than on any other porpoise or dolphin
- stocky body
- flattened forehead and no beak
- appears completely dark at sea
- little disturbance of water
- normally in small groups
- groups scatter when approached

Group size 2–3 (1–8), as many as 70 recorded together off Peru	Fin position Far behind center

Status Locally common	Population Unknown	Threats 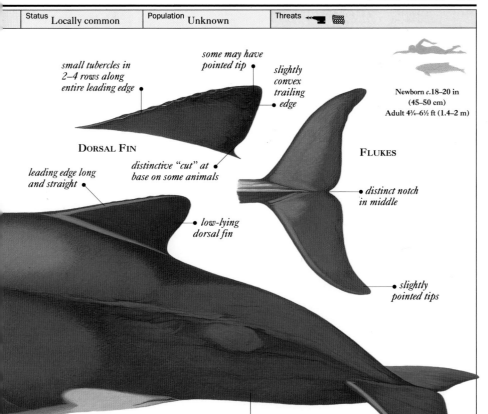

small tubercles in
2–4 rows along
entire leading edge •

some may have
pointed tip •

• slightly
convex
trailing
• edge

Newborn *c.*18–20 in
(45–50 cm)
Adult 4¾–6½ ft (1.4–2 m)

DORSAL FIN

FLUKES

leading edge long
and straight •

distinctive "cut" at •
base on some animals

• distinct notch
in middle

• low-lying
dorsal fin

• slightly
pointed tips

• thickening below
and above tail stock
(more pronounced
in older animals)

MALE/FEMALE

WHERE TO LOOK

Distribution stretches from Tierra del Fuego (the southernmost tip of South America) as far north as northern Peru
on the Pacific side and southern Brazil on the Atlantic side.
This distribution may not be continuous, although there
are not enough observations to be sure. It is believed to
be more common along the Pacific coast than the Atlantic,
and may also occur around the Falkland Islands. It seems
to prefer cold, shallow waters and estuaries near the coast.
Little is known about seasonal movements.

TEMPERATE AND SUBANTARCTIC COASTAL WATERS AROUND
SOUTH AMERICA

Birth wt Unknown	Adult wt *c.*90–155 lb (40–70 kg)	Diet

| Family PHOCOENIDAE | Species Phocoenoides dalli | Habitat 🐋 〰️ |

DALL'S PORPOISE

Dall's Porpoise rises to the surface at high speed and is usually seen as a blur. It is instantly recognizable from far away by a distinctive spray of water known as the "rooster tail"; in rough seas, however, this is hard to spot among the whitecaps. The spray is caused by a cone of water coming off the porpoise's head as it rises to breathe. The Pacific White-sided Dolphin (p.218) sometimes creates a similar spray, but it has a tall, falcate dorsal fin and more complex coloration, and it is much more acrobatic than Dall's Porpoise. There are 2 distinct forms: the Dalli-type and the Truei-type. These are distinguished by a different distribution of black and white coloring, and by size. There are also many variations in between – all-black, all-white, and piebald animals have been seen.

• **OTHER NAMES** Spray Porpoise, True's Porpoise, White-flanked Porpoise.

hooked tip (variable)

prominent dorsal fin gray-white above and black below (variable)

steeply sloping forehead

narrow mouth, with gum teeth between normal teeth

"lips" may be black or white

TEETH $\frac{38-58}{38-58}$

FRONT   SIDE

small flippers close to head

white coloring begins some way behind flippers

BEHAVIOR
Almost hyperactive. Darts and zig-zags around at great speed, and may disappear suddenly. Swimming speeds can reach 35 mph (55 km/h). Only porpoise that will rush to a boat to bow-ride, but soon loses interest in anything that travels at less than 12 mph (20 km/h). Will also ride the stern waves of a boat. Rarely leaps out of the water. Does not porpoise like other small cetaceans, but does produces a "rooster tail."

slightly longer, slimmer body than Dalli-type

MALE (TRUEI-TYPE)

white coloring begins just ahead of flippers

| Group size 10–20 (1–20), hundreds may gather at good feeding grounds | Fin position Slightly forward of center |

Status Locally common	Population Unknown	Threats

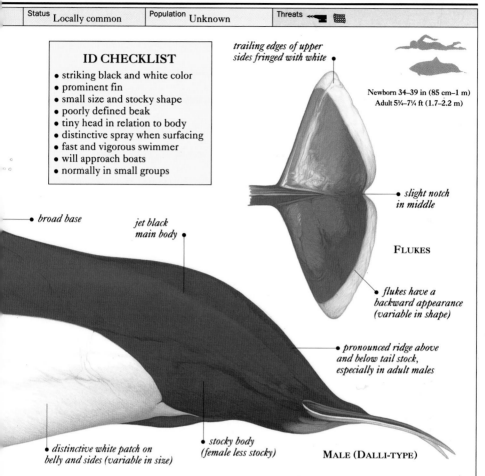

ID CHECKLIST

- striking black and white color
- prominent fin
- small size and stocky shape
- poorly defined beak
- tiny head in relation to body
- distinctive spray when surfacing
- fast and vigorous swimmer
- will approach boats
- normally in small groups

trailing edges of upper sides fringed with white •

Newborn 34–39 in (85 cm–1 m)
Adult 5¾–7¼ ft (1.7–2.2 m)

• *broad base*

jet black main body •

• *slight notch in middle*

FLUKES

• *flukes have a backward appearance (variable in shape)*

• *pronounced ridge above and below tail stock, especially in adult males*

• *distinctive white patch on belly and sides (variable in size)*

• *stocky body (female less stocky)*

MALE (DALLI-TYPE)

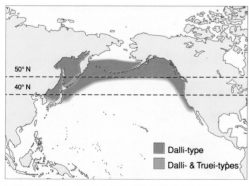

Dalli-type

Dalli- & Truei-types

BOTH EAST AND WEST SIDES OF THE NORTHERN NORTH
PACIFIC, AND IN THE OPEN SEA

WHERE TO LOOK

Occurs in the cold waters of the northern
North Pacific. Commonly found close to
land (usually near deep-water canyons),
but also in the open sea. May routinely
forage at depths of 1,640 ft (500 m) or
more. Often associates with Pacific White-
sided Dolphins (from 50° N southward)
and Long-finned Pilot Whales (from
40° N southward). Little is known about
migrations, but seems to migrate north in
summer and south in winter in the western
Pacific; in the eastern Pacific, there is a
possible inshore–offshore migration in
some areas. Some animals may summer
as far north as the Bering Strait.

Birth wt Unknown	Adult wt 300–485 lb (135–220 kg)	Diet

GLOSSARY

WORDS PRINTED in **bold** type have their own definition elsewhere in the glossary.

• **ADULT**
Sexually mature animal that is (or is almost) fully grown.

• **AMPHIPOD**
Shrimplike crustacean that is a food source for some **whales**.

• **ANCHOR PATCH**
Variable gray-white anchor or W-shaped patch on the chests of some smaller **toothed whales**.

• **ANTARCTIC CONVERGENCE**
Natural boundary in the oceans around Antarctica, where cold waters from the south sink below warmer waters from the north; lying roughly between 50°–60° S, it shifts slightly with the seasons.

• **ANTERIOR**
Situated at or near the head.

• **BALEEN/BALEEN PLATES**
Comblike plates hanging from the upper jaw of many large **whales**, used to strain small prey from seawater (also known as "whalebone").

• **BALEEN WHALE**
Suborder of **whales** with **baleen plates** instead of teeth; scientific term Mysticeti, from the Greek *mystax*, meaning "mustache," and *cetus*, meaning "whale."

• **BEACH-RUBBING**
Rubbing the body on stones in shallow water near the shore.

• **BEAK**
Forward-projecting jaws of a **cetacean** (also known as "snout").

• **BENTHIC**
Relating to the bottom of the sea.

• **BLAZE**
Light streaking of color, usually starting below the **dorsal fin** and pointing up into the **cape**.

• **BLOW**
Cloud of moisture-laden air exhaled by **cetaceans** (also known as "spout"); may be used to describe the act of breathing.

• **BLOWHOLE**
Nostril(s) on the top of the head.

• **BLUBBER**
Insulating layer of fat beneath the skin of most marine mammals.

• **BOW-RIDING**
Riding on the pressure wave in front of a ship or large **whale**.

• **BREACHING**
Act of leaping completely out of the water (or almost completely) and landing back with a splash.

• **BULL**
Adult male **whale**.

• **CALF**
Baby **cetacean** that is still being nursed by its mother.

• **CALLOSITY**
Area of roughened skin or horny growth on the head of a right whale.

• **CAPE**
Darker region on the back of many **cetaceans** around the **dorsal fin**.

• **CETACEAN**
Marine mammal belonging to the order Cetacea, which includes all **whales, dolphins,** and **porpoises**.

• **CIRCUMPOLAR**
Ranging around either pole.

• **CONTINENTAL SHELF**
Area of sea floor adjacent to a continent, sloping gently to a depth of about 655 ft (200 m); beyond the shelf edge, the sea floor drops steeply (via the continental slope) to the ocean bottom.

• **COW**
Adult female **whale**.

• **DOLPHIN**
Relatively small **cetacean** in any of several different families, with conical-shaped teeth and (usually) a **falcate dorsal fin**; as a general term, may be used interchangeably with "**porpoise**."

• **DORSAL**
Toward the upper side.

• **DORSAL FIN**
Raised structure on the back of most **cetaceans**.

• **DORSAL RIDGE**
Hump or ridge that replaces a **dorsal fin** in some **cetaceans**.

• **ECHOLOCATION**
System used by many **cetaceans** to orientate, navigate, and find food by sending out sounds and interpreting the returning echoes.

• **FALCATE**
Sickle-shaped and curved backward.

• **FLIPPER**
Paddle-shaped front limb of a **cetacean** (sometimes known as "pectoral fin").

• **FLIPPER-SLAPPING**
Raising a **flipper** out of the water and slapping it onto the surface.

• **FLUKES**
Horizontally flattened tail of **cetaceans** (containing no bone).

• **FLUKING**
Act of raising the **flukes** into the air upon diving.

• **GUM TEETH**
Horny protuberances on the gums of Dall's Porpoise, forming a tough ridge between the real teeth.

• **HERD**
Coordinated group of **cetaceans**; term often used in connection with larger **baleen whales**.

• **JUVENILE**
Young **cetacean** that is no longer being nursed by its mother but is not yet sexually mature.

• **KEEL**
Distinctive bulge on the **tail stock** near the **flukes**; it can be on the upper side, underside, or both.

• **KRILL**
Small, shrimplike crustaceans that form the major food of many **baleen whales**. There are more than 80 **species**.

• **LOBTAILING**
Forceful slapping of the **flukes** against the water while most of the animal lies just under the surface. Also known as "tail-slapping."

• **LOCALLY COMMON**
Uncommon or absent over most of range, but relatively common in one or more specific localities.

• **LOGGING**
Lying still at or near the surface.

• **MELON**
Bulbous forehead of many **toothed whales, dolphins,** and **porpoises**; believed to be used to focus sounds for **echolocation**.

• **MIGRATION**
Regular journeys of animals between one region and another, usually associated with seasonal climatic changes or breeding and feeding cycles.

• **MYSTICETI**
See **Baleen whale**.

• **NERITIC**
Pertaining to the near-shore, shallow-water zone of sea over the **continental shelf**.

• **OCEANIC**
Anywhere in the ocean beyond the edge of the **continental shelf**, usually where the water is deeper than 655 ft (200 m).

- **ODONTOCETI**
See Toothed whale.
- **PACK ICE**
Mass of floating pieces of ice driven together to form a solid layer.
- **PANTROPICAL**
Occurring globally between the tropics of Cancer and Capricorn.
- **PARASITE**
Organism that benefits from another organism while harming it.
- **PECTORAL FIN**
See Flipper.
- **PEDUNCLE-SLAPPING**
Act of throwing the rear portion of the body out of the water and slapping it sideways onto the surface, or on top of another whale (also known as "tail-breaching").
- **PELAGIC**
Living in the upper waters of the open sea far from land.
- **PERMANENT ICE**
Core areas of ice around both poles; this ice does not melt, but is surrounded by outer zones of ice that form each autumn and disperse each spring.
- **POD**
Coordinated group of cetaceans; term often used in connection with larger toothed whales.
- **POLAR**
Of the areas around the poles.
- **POPULATION**
Group of animals of the same species that is isolated from other such groups and interbreeds.
- **PORPOISE**
Small cetacean in the family Phocoenidae, with an indistinct beak or no beak, a stocky body, and spade-shaped teeth; most have a triangular dorsal fin; as a general term, may be used interchangeably with "dolphin."
- **PORPOISING**
Leaping out of the water while moving forward at speed.
- **POSTERIOR**
Situated at or near the tail.
- **PURSE-SEINING**
Fishing with a long net – up to 1¼ miles (2 km) in length and 330 ft (100 m) deep – that is set around a shoal of fish to form a circular wall, then gathered at the bottom and drawn in to form a "purse."
- **RACE**
Interbreeding group of animals that is genetically distinct from other such groups of the same species; races are usually geographically isolated from one another.

- **RANGE**
Natural distribution of a species, including migratory pathways and seasonal haunts.
- **RESIDENT**
Stays in one area all year round.
- **ROOSTER TAIL**
Spray of water formed when certain small cetaceans surface at high speed; it is caused by a cone of water coming off the animal's head.
- **RORQUAL**
Strictly speaking, a baleen whale of the genus *Balaenoptera*; however, many experts also include the Humpback Whale (genus *Megaptera*) in this group.
- **ROSTRUM**
Upper jaw of the skull (may be used to refer to the beak or snout).
- **SADDLE PATCH**
Light patch behind the dorsal fin on some cetaceans.
- **SCHOOL**
Coordinated group of cetaceans; term often used for dolphins.
- **SEAMOUNT**
Isolated undersea mountain (usually a volcano) with the summit lying well below the ocean surface.
- **SNOUT**
See Beak.
- **SONAR**
System used by many cetaceans to echolocate.
- **SOUNDING DIVE**
Deep (and usually longer) dive after a series of shallow dives (also known as "terminal dive").
- **SPECIES**
Group of similar animals, reproductively isolated from all other such groups and able to breed and produce viable offspring.
- **SPLASHGUARD**
Elevated area in front of the blowholes of many large whales, which prevents water from pouring in during respiration (also known as "blowhole crest").
- **SPOUT**
See Blow.
- **SPYHOPPING**
Raising the head vertically out of water, then sinking below the surface without much splash.
- **STRANDING**
Act of a cetacean coming onto land, either alive or dead; mass stranding involves a group of 3 or more animals.
- **SUBMARINE CANYON**
Deep, steep-sided valley in the continental shelf.

- **SUBSPECIES**
Recognizable subpopulation of a species, typically with a distinct geographical distribution.
- **TAIL STOCK**
Region from just behind the dorsal fin to the flukes (also called "peduncle" or "caudal peduncle").
- **TEMPERATE**
Mid-latitude regions between the tropics and the polar circles, with a mild, seasonally changing climate; cold temperate regions are toward the poles, warm temperate regions are nearer the tropics.
- **THROAT GROOVES**
Grooves on the throat present in some groups of whales.
- **TOOTHED WHALE**
Suborder of cetaceans with teeth; scientific term Odontoceti, from the Greek *odous*, meaning "tooth," and *cetus*, meaning "whale."
- **TRANSIENT**
Always on the move rather than staying mostly in one area; usually refers to Killer Whales.
- **TROPICAL**
Pertaining to low latitudes of the world between the tropic of Cancer and tropic of Capricorn.
- **TUBERCLES**
Circular bumps along the edges of the flippers and dorsal fins of some cetaceans; also the knobs on a Humpback Whale's head.
- **VENTRAL**
Relating to the underside.
- **VESTIGIAL**
Pertaining to part of an animal that is in the process of being evolutionarily lost and is small, imperfectly formed, and serves no function.
- **WAKE-RIDING**
Swimming in the frothy wake of a boat or ship.
- **WEST-WIND DRIFT**
Principal circumpolar current around Antarctica, flowing in an easterly direction.
- **WHALE**
General name applied to any large cetacean and a specific name applied to certain smaller ones.
- **WHALEBONE**
See Baleen/baleen plates.
- **WHALE LICE**
Small, crablike parasites that live on some species of whale.
- **WHALING**
The intentional hunting and killing of whales for their meat, blubber, baleen, and other products.

FURTHER READING

Bryden, M.M. and Harrison, R.J. (Eds.) *Whales, Dolphins and Porpoises*, Merehurst Press, London, 1988

Carwardine, Mark *On the Trail of the Whale*, Thunder Bay Publishing Co., UK, 1994

Evans, Peter *Whales*, Whittet Books, London, 1990

Hoyt, Erich *The Whale Watcher's Handbook*, Doubleday, New York, 1984

Klinowska, M. *Dolphins, Porpoises and Whales of the World*, The IUCN Red Data Book, IUCN, Cambridge (UK), 1991

Leatherwood, Stephen and Reeves, Randall *The Sierra Club Handbook of Whales and Dolphins*, Sierra Club Books, San Francisco, 1983

Martin, Anthony R. *Whales and Dolphins*, Salamander Books, London and New York, 1990

May, John (Ed.) *The Greenpeace Book of Dolphins*, Century Editions, London, 1990

Obee, Bruce and Ellis, Graeme *Guardians of the Whales*, Whitecap Books, Vancouver and Toronto, 1992

Ridgeway, S.H. and Harrison, R. (Eds.) *Handbook of Marine Mammals: Vol III, 1985; vol IV, 1989; vol V, 1994*, Academic Press, London et al